Psychogenesis and the History of Science

PSYCHOGENESIS AND THE HISTORY OF SCIENCE

Jean Piaget
Rolando Garcia

Translated by
Helga Feider

COLUMBIA UNIVERSITY PRESS
NEW YORK

LIBRARY OF CONGRESS
Library of Congress Cataloging-in-Publication Data
Piaget, Jean, 1896–
[Psychogenèse et histoire des sciences. English]
Psychogenesis and the history of science / Jean Piaget, Rolando Garcia ; translated by Helga Feider.
p. cm.
Translation of: Psychogenèse et histoire des sciences.
Bibliography: p.
Includes index.
ISBN 0-231-05992-2 (alk. paper)
1. Cognition—History. 2. Knowledge, Theory of—Psychological aspects. 3. Science—Psychological aspects. 4. Science—History. I. Garcia, Rolando. II. Title.
BF311.P517713 1988
153—dc19 88-9578
CIP

Columbia University Press
New York Guildford, Surrey

Printed in the United States of America

Casebound editions of Columbia University Press books are Smyth-sewn and are printed on permanent and durable acid-free paper.

Book design by Ken Venezio

Contents

Foreword

The two authors have been kind enough to ask me to write a few words of introduction to their book, which is the culmination of a unique collaboration between two epistemologists, one a psychologist, the other a physicist. Piaget had the great joy to see it completed the day before he contracted the illness that ended his days.

I was privileged to have witnessed, from 1967 on, the constant exchanges during the conception and writing of *Psychogenesis and the History of Science*. They are reflected in the different chapters, which bear the personal stamps of the two authors. I note here the degree to which the thinking of both Piaget and Garcia has been enriched and modified by this mutual contact.

Piaget dedicated himself to the creation of a genetic epistemology. He appropriated to himself a historical-critical method based in turn upon a psychogenetic method. His vision of the growth of knowledge in the child, refined and deepened by the historical study of scientific thought, continued to renew itself, even to the present work.

In similar fashion, Rolando Garcia, a student of Carnap and Reichenbach, discovered, thanks to Piaget, the rich and revealing facts of the psychogenesis of the representations that the child has of the universe. Garcia has been led to envision in a new way the evolution of scientific thought from Greek antiquity to the Newtonian revolution. In this volume, he analyzes a few chapters of that history, in which the mechanisms of progress are particularly evident.

The two authors have not confined their interest to an examination of their parallel fields of research. In trying to discover the processes involved in all construction of knowledge, each has subordinated his respective field of psychogenesis and history of science in order to verify the hypothesis of a constructivist epistemology.

The present work constitutes, among the writings of Piaget and his co-workers, the third and, it seems to me, most important epistemological synthesis.

In his first synthesis, *Introduction to Genetic Epistemology* (1950),[1] which at the time, was already a summation, Piaget interpreted the results of his research on the psychogenesis of certain categories of thinking (number, physical quantity, space, time, etc.) in the light of classical epistemologies. He showed that both aprioristic idealism and empiricist realism are insufficient. He stressed that to give epistemology a scientific status it is important to study the processes that transform knowledge, be they ontogenetic or historical: "the nature of a living reality does not reveal itself either in its initial or final stages, but in the very process of its transformations: this is the law of construction, that is, the operational system in its progressive constitution." Only the growth of knowledge will permit us to determine the respective contributions of the subject and the object.

It is clear from his first introduction to the discipline that would become the foundation of his entire work that Piaget believes that this discipline "should remain an open science." Furthermore, Piaget was convinced that an epistemology which aims to be scientific—that is, communicable independently of any particular school of thought—can only be the fruit of interdisciplinary efforts. He was therefore happy that the great logician and mathematician Evert W. Beth (who died prematurely) agreed to collaborate with him, even after he had severely criticized Piaget's essays on formalization, and to be co-author of *Epistémologie mathématique et psychologie. Essai sur la logique formelle et la pensée réelle*. (1961).[2]

In this work, which represents the second synthesis of epistemology as conceived by Piaget, Beth analyzed the foundations of mathematics while defending the principle that logic and the psychology of knowledge should be autonomous diciplines. He aligned himself with Piaget's position, according to which epistemology explains how man's empirical thought can produce science conceived as a coherent system of objective knowledge. Piaget, in turn, seeks to show that cognitive structures derive from more general systems of coordination of actions. The two authors came to the conclusion that the norms elaborated by the epistemic subject in the course of her development are comparable to those inherent in scientific thought. This book, thus, constitutes a provisional completion of an attempt to relate formal models to empirical thinking, which characterized the "structuralist" period in Piaget's research.

With this new impetus, Piaget embarked upon the third phase of his research by studying two important problems that had remained unresolved until then: physical thought and a complete reorganization of a theory concerning developmental mechanisms.

Research on theories of causality[3] has brought together such theoreticians as the late Rosenfeld, Bunge, Kuhn, Halbwachs, and, somewhat later, Garcia. This work stimulated a great deal of psychological research, of which a particularly noteworthy outcome was Piaget's and Garcia's monograph *Les explications causales*.[4]

This work on the ontogenesis of causal explanations led Piaget to focus on the role of the object in the development of operational thinking and to develop an integrated interactionist approach. The genetic epistemology of the categories of knowledge was completed, the logico-mathematical models that served as structural tools in their analysis had been constructed. This was the time to analyze in greater depth the very mechanisms that mediate the growth of knowledge in the child. Their relevance to the comparative study of the history of science will become evident.

These mechanisms can be grasped when one examines the transitions leading from a level of organization where the subject is well adapted to the environment to those levels where such adaptation is relatively poorer. Piaget has tried to account for this improvement or augmentation of knowledge in terms of a model borrowed from biology: that of equilibration. This internally regulated mechanism neutralizes disturbances that produce states of disequilibrium. However, re-equilibrations do not necessarily produce homeostasis, but can be the source of innovation. There are many specific processes that determine the changes from one level of organization to the next and the integration of prior forms of knowledge within the new ones: reflective abstraction and generalization, the process of conscious appropriation, and thematization, the invention of possibilities and inferencing, which leads to necessity. All these determine an evolution, which proceeds in spiral fashion, and is dialectic in nature, indicating the important role desequilibria have in stimulating the subject toward productive re-adaptations.

These mechanisms and instruments of progress have been illustrated by the great number of behaviors children adopt in solving problems. They turned out to be of a more general nature so that they can serve as a heuristic for analyses in depth of the historical sequences in the evolution of certain aspects of mathematical and physical thinking. The authors' purpose in studying these generalized mechanisms was not to describe term-by-term correspondences, and even less to propose a recapitulation of phylogenesis by ontogenesis, nor even to demonstrate the existance of analogies in sequencing. Instead, they wished to see if the mechanisms mediating the transition from one historical period to the next, for particular notional systems, are analogous to those mediating transitions from one developmental stage to the next. Clearly, the most salient examples are the successive explanations children give of the transmission of motion. These are worked out as a function of operational development. They are comparable to the way impetus was explained at different historical periods by different philosophers and scientists, from Aristotle to Buridan and Benedetti. The study is in essence directed toward the question of why human thinking proceeds in sequential progressions of this kind. Epistemologically more important and unexpected is the discovery of a general process, which leads from an analysis of objects (intra-object) to one concerning relations of transformations between objects (which one might call "inter-object") to a final level of analysis, which the authors call "transobjectal," and which concerns the building of structures.

In the history of geometry, Garcia distinguishes three periods: (a) the geometry of Greek thinking until the eighteenth century; (b) the projective geometry of Poncelet and Chasles, and (c) the global conception of geometry introduced by Klein. The descriptive geometry of Descartes and Fermat, and differential and integral calculus provide the instruments which ensure the transition from (a) to (b), and the theory of groups the transition from (b) to (c). The similarities between such progressions noted over the centuries and the spatial and geometric representations of the young child, which proceed from topological intuitions to the construction of abstract systems of reference after going through the elaboration of projective notions, pose intriguing questions for a constructivist epistemology.

The sequential relationships of human discoveries and knowledge—whether in the fields of algebra and geometry or in the various domains of physics, particularly mechanics—as observed and presented by Piaget and Garcia, continue to pose new problems for epistemology. Thus, thanks to the productive collaboration of these two great scholars, their masterful research opens up new directions for scholarship that will extend the limits of psychology and the history of science.

Bärbel Inhelder

Acknowledgments

I would like to thank particularly Bärbel Inhelder, witness and participant to several of the discussions on the topics taken up in this volume. She has accepted the task of editing the text of the chapters written by Jean Piaget (assisted in this by Denys de Caprona and Marguerite Ravex-Nicolas). I also wish to thank Hermine Sinclair and Pierre Spitz, who read the manuscript and made many suggestions toward improving the style of the chapters for which I was responsible. This work could not have been realised without the collaboration of my wife, Emilia Ferreiro.

Initial drafts of this book were prepared as early as 1974, but the final version was not ready until 1980. Additional analyses and discussions of the manuscript took place at different times in the elaboration of the text (I gratefully remember intense and fruitful discussions during excursions to the mountains.) Since, during all that time, epistemology was far from being my only pursuit, it took longer than expected to complete the manuscript. I must give recognition to Jean Piaget for his kindness and tolerance in accepting my own "working times," in spite of his urgent need to finish this work, the publication of which he had hoped for and announced on several occasions.

The research that gave rise to this book was carried out at the International Center of Genetic Epistemology at Geneva and was funded by the Fonds national suisse de la recherche scientifique and by the Ford Foundation.

R. G.

Psychogenesis and the History of Science

Introduction

Among scholars and historians of science, it is commonly believed that there is no relation between the way concepts and operations are formed in their elementary stages and the way they evolve at higher levels. There is also a frequent though less general belief that the epistemological significance of a conceptual tool is independent of its mode of construction. It is believed that the mode of construction concerns history, and perhaps psychogenesis, while the epistemological significance pertains to the way the instrument functions within an actual, synchronic system of cognitive interactions, which, according to that hypothesis, are irreducible to diachronic considerations.

I. LEVELS OF DEVELOPMENT

This generally low level of interest in elementary stages is undoubtedly due in part to the contemporary conception of the development of knowledge as a linear process, wherein each stage replaces the preceding one, with which it usually conserves certain ties. The process however, is not believed to extend to the elementary levels. The truth is that, the stages in the construction of different forms of knowledge are actually sequential—so that each stage is at once the result of possibilities opened up by the preceding stage and a necessary condition for the following one.

In addition, each stage begins with a reorganization, at another level, of the principal acquisitions that occurred at the preceding stages. As a result, even at the highest levels there is integration of certain associations the nature of which can be explained only by analyzing the early stages. This book presents many examples of this process. But to give a rough idea of what is meant, let us present here—very briefly—one or two of them.

For sequential order in history, one might cite the three great periods in the evolution of mathematics: the static realism of the Greeks, which concerns only permanent states (figures and numbers). Nevertheless it provided a certain knowledge base necessary for the discovery of the algebraic and infinitesimal transformations of the seventeenth century. The analysis of the latter was, in turn, indispensable for the constitution of the structures proper to the mathematics of the nineteenth century—and of today. It is clear that in physics, the discovery of new facts can modify the course of ideas in a variety of ways, so that this science cannot be expected to present the same kinds of trends as the field of mathematical logic.

As for reorganization from one level to the next, with integration of elements going back all the way to the initial phases, it is useful to distinguish empirical abstraction, which draws its information from the objects themselves, from what we call "reflective abstraction."[1] The latter proceeds from the subject's actions and operations, according to two processes that are necessarily associated: (1) a projection onto a higher level (for example, of representation) of what is derived from the lower level (for example, an action system); and (2) reflection, which reconstructs and reorganizes, within a larger system, what is transferred by projection.

This reflection is a constructive process, for two complementary reasons. First, projection essentially establishes correspondences at the next higher level, associating the old contents transferred from previous levels with new contents that can be integrated within the initial structure but permitting it to be generalized. Second, these first organizations also lead to the discovery of related contents, which may not be directly assimilated into the earlier structure. This makes it necessary to transform that structure by means of a constructive process until it becomes integrated within a larger, and therefore partially novel, structure. This mode of construction, by reflective abstraction and "constructive generalization,"[2] repeats itself indefinitely, at each successive level, so that cognitive development is the result of the interactions of a single mechanism—one, however, which is constantly renewed and extended by the addition of new contents alternating with the creation of new organizations of structures. These are the two reasons why the most advanced constructions conserve partial links with their most primitive forms: one because there are successive integrations, the other because the functional identity of an iterative mechanism constantly renews itself by being repeated at different levels.

Let us cite a general example of this process before analyzing it in detail. We may take note of the relations existing between the contents of observables and their logico-mathematical structure. The acquisition of experimental evidence proceeds certainly in successive approximations, related to the construction of recording equipment, which, in turn, depends on theoretical models and the new problems raised by them. The result is a progressive extension of observational scales in two directions from the bottom up as well as from the top down. Naturally, necessary reorganizations of the whole occur with each new refinement. The increasingly complex levels of the mathematical formulation of observables—and particularly its considerable variability in the course of history—has led to two kinds of belief. One is well founded while the other is questionable.

What cannot be doubted is that no matter how sophisticated the level of mathematical formulation is on a scientific level for a given physical phenomenon, the latter still corresponds to a datum external to the subject; hence, objects exist, even though they can be attained only by approximations which permit nearly exact, never exhaustive descriptions, so that they will always remain in a state of limited access. More questionable is the belief that if formulation is the work of the subject and if the object exists, then it must be possible to trace a stable limit between the mathematization and the objects themselves, since a physical "fact" does not in itself include a logico-mathematical dimension, but receives it only after the event.

This is where the analysis of the most primitive reactions becomes important, since it provides a definitive answer. Not only is there an absence of a clear frontier between the contributions of the subject and those of the object (since we can only know about interactions between the two) but in addition, it is only to the extent that logical and mathematical structures are applied that one can come to attain the object, and objectivity improves as a function of richer logico-mathematical structures. In fact, the elementary perceptual object is already partially "logicized" from the start, even though much less "objective" than the more elaborated object. This initial logicization occurs because, in order to isolate objects from global scenes and to attribute permanence to these objects, it is necessary that the actions directed at the objects be coordinated in forms of assimilation, hierarchical order, correspondence etc., which are already logico-mathematical in nature. The interdependence of quantified spatial relations (as more or less) of such forms to their contents, which is attainable only within the frame of reference given by the subject's cognitive activity is thus completely general at all levels. It can be verified only by means of psychogenetic analyses.

II. FORMATION AND SIGNIFICANCE

We now come to the central problem, which will be discussed and rediscussed in the present volume. Is the formation of cognitive tools such that it can shed light on their epistemological significance? Or do the two belong to entirely different domains—the former to psychology and history, the latter to a realm that calls for methods that are radically independent from the former?

But the term psychogenesis gives rise to troublesome and very tenacious misunderstandings, as long as the following two problems and fields of study are not systematically distinguished: (1) the psychogenesis of knowledge, or the study of the formation and nature of the cognitive tools, are subordinated to the norms which the subject sets or accepts for herself in her endogenous intellectual activities or activities that concern objects; (2) the psychogenesis of factual processes, independent of any kind of norms—that is, independent in terms of true and false (from the subject's point of view). This concerns only the way behaviors function on a psychophysiological level (material mechanisms of action, states of consciousness, memory, mental images, etc.).

It is clear that authors who contest the relevance of psychogenesis for epistemology see only this factual aspect of development and forget that, at all levels, subjects obey certain cognitive norms. The latter, however, are of interest because of the dynamics of their successive constructions, of unsuspected scope, and necessary in all constitution of valid knowledge. It is true that we are dealing here only with pre-scientific norms, but the fact of fundamental importance for epistemology is that the subject, beginning with very low level prelogical structures, comes to develop rational norms that are isomorphic with those of the early days of science. To understand the mechanism of this development of prescientific norms up to their fusion with those of nascent sciences, scientific thought is undeniably an epistemological problem. It has, incidentally, been treated quite often on the level of the sociogenesis of knowledge, with respect to "natural" numbers (a term which in itself is the formulation of a problem and even of our problem), with respect to classification, etc. The logician E. W. Beth, who can hardly be accused of being exceedingly enamored of psychology, after having reaffirmed the necessary autonomy of logic (with no psychologism) and of the psychology of knowledge (without "logicism," since the norms to be studied are those of the subject, not those of the logician), adds the following remark: "But the situation is quite different, if one places oneself at the point of view of the epistemologist to the extent that this discipline has for its object to interpret science as a result of human mental activity or, which amounts to the same, to explain how human thinking about reality can produce science defined as coherent systems of objective knowledge."

As for specifying how the study of the subject's cognitive norms can help in understanding the processes involved in the constitution of knowledge, it is obvious that we cannot do so by resorting to verbal declarations or even by analyzing the process of growth of consciousness, but in essence by examining what the subject "does" (as opposed to what she thinks she does) in order to acquire and make use of a given type of knowedge or skill or to make sure of its being reasonably well founded. It can, thus, happen that the same problem presents itself in the domain of the epistemology of science and that of developmental psychology. A good case in point is that of the notion of the undeformable (rigid) solid, which many scholars of geometry in the nineteenth century considered as an empirical, and even perceptual datum, and this led them to see their science as mathematics applied to experience.

In contrast, it has since been shown that to attribute to a solid this (ideally) undeformable characteristic in fact depends upon complex deductive structuring, the group of displacements, and a metric conserving the invariance of distances. A recent experiment[4] at the International Center for Genetic Epistemology presented children with a triangular device, made out of stiff rods connecting three small rings, *A*, *B*, *C*. Then, attaching a pencil to *A*, one performs various translations, such as — or |, L, or Z, etc. The subject's task is then to reconstruct the paths followed by *B* and *C* on the same paper where the path of *A* had just been traced by placing *B* and *C* under *A* in such a way that the three points constitute a rigid triangle which cannot be put out of shape, with *A* as a vertex.

The youngest subjects cannot even conserve the initial configuration at the endpoint. At a later age, they respect this correspondence, but seem unconcerned about the intermediate positions, as if *A*, *B*, and *C* were able to move independently of each other in the process, while returning to their original relative positions again at the end. Only at a fairly late stage, and then only on the basis of metric arguments and a postulate of invariance of particular relations in the course of displacements, do subjects succeed in solving this problem.

Evidently, the indeformability of the device had to be constructed and was not given by visual perception, which was constantly available, while the global object was slowly rotated. We have here an example of an epistemological question (whether the concept of an indeformable solid is given empirically or constructed by a deductive process), which presents itself both at the level of the philosophy of geometrical knowledge and of psychogenesis. It can be noted that the answer given by the latter confirms the rationalist interpretation, while if the empiricist position were correct, one would have found such confirmation at the elementary level.

III. PROBLEMS OF HISTORY

As to the question whether the formation of knowledge can provide information about its epistemological significance, on the level of the history of science itself and no longer that of psychogenesis, this hypothesis is more easily and more generally accepted. In fact, it is now generally known that science is in perpetual flux and that there is no field, however limited, that can be considered to be definitely established in its bases and sheltered from any future modification, even though, as in mathematics, what has been demonstrated is integrated in what follows and is not questioned again. Such integration can, in effect, reveal that a truth that was considered general has actually constituted only a special case. When this happens, one can speak, in this restricted sense, of a partial error and of rectification.

Given these conditions of general change, it is not surprising that a piece of knowledge cannot be dissociated from its historical context and that, consequently, the history of a concept gives some indication as to its epistemic significance. Still, to understand this relationship, it is necessary to pose the problems in terms of "trends"—that is, in terms of the evolution of norms at a scale that makes it possible to discriminate stages, rather than in factual terms of how one author influences another; or, particularly, in terms of the controversial if somewhat uninteresting problem concerning the role of precursors in the creation of a new system. The essential problem is how to characterize the important stages in the evolution of a concept or a structure or even of the general perspective concerning a particular discipline, irrespective of accelerations or regressions, the impact of precursors or "epistemological gaps." The central problem, in fact, is not that of continuities or discontinuities (both play a role in any kind of development),[5] but rather that of the existence of the stages themselves, and particularly that of explaining their sequence. As for the role of precursors, this is a psychological problem much more than an epistemological one, depending on whether the later author wishes to extend or complete the intuitions of his predecessors or instead contest them. Finally—and this is the most frequent situation—he may waver between the two attitudes.

Now, if it is true that everyone immediately recognizes the epistemological relevance of studying the historical periods and their directions, it would be difficult to imagine that someone would find it illogical to recognize the informative value of history in analyzing the construction of knowledge, while contesting it with respect to psychogenesis. But the relationship between the two kinds of research is close, not only because the stages of knowledge, as we have just seen, do not simply follow each other in a linear sequence, in which case one might expect that the early stages play no role whatever later on, but also because each stage or period begins with a reorganization of what it has inherited from preceding ones. This explains the partial integrations up to the higher levels of some of the initial positions. We cited an example of this in section II, concerning the notion of a nondeformable solid. In that example, the empirical nature of the concept would have been confirmed if it had been verified psychologically, even to the point where it had epistemological significance. Instead, the deductive, constructive nature of the concept became evident, since this is the way that the property of nondeformability is, in fact, constructed.

The main reason why there is a kinship between historico-critical and genetic epistemology is that the two kinds of analyses, irrespective of the important differences between them in the data used, will always, and at all levels, converge toward similar problems as to mechanisms and instruments (e.g., reflective abstractions, etc.). These mechanisms operate not only in the elementary interactions between subjects and objects, but particularly in the way that a lower level influences the formation of the following one. This inevitably leads to a situation where, as will be seen, the same general problems, common to all epistemic development are posed.

IV. EXPERIENCE AND DEDUCTION

The most general problem is, of course, the one raised by the example just cited, but which can be found in all domains: determining the relative contributions of experience and the subject's operational structures in the elaboration of knowledge. To cite just one case, the problem of principles of conservation, it is almost trivial to bring up the following two kinds of considerations. On the one hand, conservation is a requirement inherent in the notion of deduction itself, since, if every thing changes at once and nothing stays invariant, no necessary inference can ever be made. This is why, before their measurement, Lavoisier postulated the conservation of weight in chemical reactions. This is also why the notion of potential energy imposes itself independently of any direct measure. On the other hand, the principles of conservation have changed in content, as has been the case with mass and energy in relativity theory, or with the "actron" in microphysics.

This shows that experience also plays an indispensable role. Hence Poincaré's remark: we know that something is always conserved, but only experience tells us what it is. In this respect, one of our most interesting findings was that even the most basic conservations are not immediate, like those related to simple additivity. We were then able to observe, step-by-step, the formation of the operations that constitute the invariants.

Concerning the first point, one might cite the initial failure to conserve number,[6] when one changes only the spatial disposition of seven to ten objects. Up to the age of seven years, children believe that the number increases when the elements are more widely spaced, so that the length of the row is modified. On the second point, let me briefly describe what happens when a sausage-shaped piece of clay is transformed into a ball.[7] After first denying conservation of substance, weight, and volume, subjects then come to believe that the quantity (or total substance) necessarily remains the same, while weight and volume are still believed to have been modified. But what would a quantity of substance be, independent of its weight and volume? Certainly, quantity is not a perceptual datum; it is not even perceptible, but a logical requirement. In addition, it is possible to observe how it develops. At the level of nonconservation, subjects, of course, do not fail to notice that the change in shape is due to changes in the position of parts, but they interpret these changes only in terms of new products appearing at the point of arrival, not thinking about what disappeared at the point of departure. If instead of simply manipulating the clay with one's finger, one takes away a piece of clay from the ball and then adds it on again to make it longer, subjects accept conservation more readily.

This confirms our hypothesis. The constitutive operation in question can, thus, be characterized in terms of "commutability," understood as equalization or compensation between what is subtracted at one point and what is added at the other (independent of linear order otherwise characterizing commutativity): the generality of this inferential process seems to be of some epistemological interest, since it subsumes invariance within a system of operational compensations, which gives it its deductive character, provided that experience furnishes the apropriate, corresponding contents.

V. INITIAL INSTRUMENTS OF KNOWLEDGE

This leads us to a second important problem of cognition common to the historico-critical and genetic epistemology, that of the relations between the subject and the objects of her knowledge, whether these are logico-mathematical or physical in nature. At this point we do not yet ask this question in terms of the great traditions (the empiricist, aprioristic or dialectical, etc.). However, our inquiry will eventually lead to such interpretations. Rather, we simply wish to establish what kinds of tools subjects use to solve problems, where these tools come from, and how they become elaborated. When one considers a system of knowledge in its completed state (for example, when it has become axiomatic), one might get the impression that such systemized knowledge can be reduced to a series of statements. Logical positivism, in its beginnings had even tried to see in language and perception the origin of the instruments of all logico-mathematical and physical knowledge. However, what has to be done, of course, is to examine the tools that had been used to acquire this knowledge before these formalizations—since the latter refer necessarily to a prior acquisition—unless one locks oneself into pure logic, which, in a certain way, represents a formalization of formalizing activities.

Let us first note that, even on the level of strict axiomatics, we distinguish two processes which one might accept as common to all fields of knowledge and which can be traced in their various manifestations throughout the history of science. These are comparative tools, or setting up of correspondences, and transformational tools or operational constructions. In fact, an axiomatic system includes a set of implications, which are, for example, if $p \supset q$, the set of injective correspondences between the truth of p and that of q. Theorems, on the other hand—if they are not implied by a single axiom—result from combinations of nonredundant axioms. In this case, there are "transformations," if we designate as such the construction of new contents from others in such a way that the latter are not analytically implied by the former. In effect, the composition of distinct axioms leads to something more than an enumeration of what is given in each individual axiom. Hence the product of such a synthesis is not analytic.

Considering now the knowledge acquired before it gets formalized, we find at all levels the same twofold process in the constitution of instruments: comparative tools, built on correspondences, and transformational operations. We find, however, that the two become increasingly interrelated, particularly since it has been possible to study transformations in themselves. Ancient Greek mathematics made use of many operations, or transformations of a figure in geometrical demonstrations or in numerical manipulations. Likewise, it is obvious that any equation incorporates a series of correspondences; these had even been developed into a complete theory of proportions. But because the ancients did not consciously grasp the transformations as such, the correspondences did not attain the intertransformational level that they were to present explicitly in the algebra and infinitesimal analyses of the seventeenth century.

From that period on, we note the construction of an increasing multiplicity of transformations recognized as effective operations narrowly associated with correspondences. But the final thematization of transformations as "structures," and of correspondences as "categories," was accomplished at two different moments, which is an additional and significant proof of their duality. One of us has tried elsewhere to explain why a thematic theory of correspondences appeared later than the theory of transformations: Since all thematization is a matter of making comparisons based on correspondences, reflective comparisons on the structures of transformations are direct, hence first-order, while the thematization of categories is a second-order process, since it involves reflecting upon instruments, which are themselves reflective, hence of second degree.

We introduced these facts, because they provide a good example of historical processes that may be illuminated by psychogenetic analyses. In the first place, these analyses have shown decisively that the first instruments of knowledge are neither perception nor language, but rather sensorimotor actions organized in schemes, which dominate over perception from the beginning and only much later become verbalized as concepts and internalized as cognitive operations. Now, each action scheme is a source for correspondences to the extent that it gets applied to new situations and objects, while the coordination of the schemes is a source of transformations in that it generates new possible actions. So we see not only that the duality exists from the very beginning, but also that the two processes—correspondence and transformation—operate together.

However, it is also essential to note that conscious reflection is first oriented toward the outside (the results of action) rather than inside (endogenous coordination of actions), so that correspondences remain independent of transformations for a long time, coming to interact only later. This state of affairs parallels one in the early phases of scientific thought for the same reason: the laws that govern the development of consciousness are very general. They explain, among other things, why, at all levels, the thematization of an operation always occurs later than nonreflective use of the same operation.

VI. CONSTRUCTION AND PREFORMATION

A third important problem to which the comparison between the history of science and psychogenesis may be relevant is the question of whether new knowledge is pre-formed or arrived at by effective construction, which in turn may have been predetermined. It may appear strange to ask this question with respect to psychogenesis, since, as commonly understood, the child does not invent but receives everything she learns through education. Now, the best proof against this belief is the spectacular development one observes during the first 18 months of life, when the child is not yet able to talk and has learned only a very small number of behaviors. Still, the development of her intelligence, the construction of space, of object permanence, of causality, etc. bear witness to an astonishing multiplicity of inventions and discoveries. Thus, even at this period the problem is to determine whether these innovations follow a sequence—whether they derive from a hereditary program, or whether they represent the realization of possibilities present from the beginning in the form of a kind of *a priori* synthetic intuition.

In the domain of deductive science, the problem is very difficult. However creative an invention may have appeared at its inception, its consequences—once they have been established—will seem necessary to such a degree that one cannot imagine a world without these innovations. This occurs to such an extent that even without appealing to a transcendental or transcending subject possessing all logicomathematical truths, mathematicians are almost always more or less Platonists. But, how are these mathematical "beings" to be conceptualized and where are they to be situated? Now, the notion of these "beings" has undergone profound modifications in contemporary scientific thinking: any operation or morphism is such a "being" in the same way as a number or a figure, in which case it is difficult to imagine what could be an operation or a comparison by correspondence without a subject that operates or compares. Are we, then, dealing with a world of "possible" constructions? But the set of "all" possibilities is an antinomy and what is given consists in nothing but new avenues toward continually new possibilities. This leads us to a constructivist position, but with a new condition: with each new possibility that opens up, the question of whether or not it was predetermined presents itself again.

There are, in fact, only two coherent positions one can take. On the one hand, assume that the mathematical "beings" exist outside of nature and, with G. Juvet,[9] see nature as only a very small sub-sector of the universe of these "beings." On the other hand, one can consider these beings as a part of nature, and subject to temporal processes: they become atemporal and permanent once their intrinsic necessities are constituted. But having decided that the logicomathematical forms are a part of human cognition, one can still ask whether they are predetermined by a hereditary program or result from a sequence of constructions, each of which presents an innovation.

It is at this point that a recourse to history and psychogenesis becomes mandatory: the former unambiguously shows that necessity evolves (what has remained today of what seemed "necessary" to the Greeks?), while the latter can inform us about the way it becomes constituted in its simplest forms. Now, since we have to choose between innateness and constructivism and, given that the human mind is a product of biological organization before the cultural community comes to predominate, it would seem that the fundamental organic mechanism in development is not heredity, but rather a system of self-regulations. To consider mathematical structures as innate would raise serious biological problems, and it would hardly dispel any of the mysteries of their nature, while the mechanism of regulations—with its compositions and retroactive feedbacks—appears already in itself to prepare the operational, reversible structures. On the other hand, the long and laborious series of trial-and-error procedures, which in psychogenesis precede the construction of invariants, of transitivity, and of recursivity itself would be difficult to understand, if the fundamental operational properties were genetically programmed, and their relatively very late emergence (around seven years of age) would have to be attributed only to slow maturation of the nervous system.

This seems all the less plausible in that it is possible, by changing experiential procedures, to accelerate or to slow down such constructions. Now, once these become accessible, they tend to be followed, more or less readily, by generalizations which sometimes go to infinity. A five-year-old we observed accepted that if $n = n$ then $n + 1 = n' + 1$. When we asked what would happen if one went on like this for a very long time, the child replied: "If one knows it one time, one knows it for all time."[10]

It is this transition from a temporal construction to an atemporal necessity that is easy to explain within a constructivist perspective, but which would pose the insolvable problem of where to localize the infinite, if it were assumed not to lie in the unlimited iteration of an "action," but to consist in a genetically predetermined "being" or in a platonic kind of reality, of which no one has ever been able to say by what mechanism it can be attained.

All the facts we shall describe, showing the analogies existing between the historical constructions and psychogenetic processes, will confirm an already well known truth that is illuminating as to the choice between the constructivist and the preformation hypothesis: before a structure becomes thematized, the operations which constitute it are already operative during the phase preceding its construction, so that the construction is built on a structure in the process of development. In the same way, a developed construction will require integration within a stronger, more advanced structure. These facts,which concern both temporal construction and the atemporal theory of the limits of formalization, show the relatedness of the two kinds of apparently heterogeneous considerations and the necessity to conceive of all logicomathematical beings as owing their existence only to their own acts and not to a datum independent of the subjects.

VII. DIFFERENT PATHS AND DIFFERENT RESULTS

But, if the mathematical "beings" are generated by the subject's constructions while truths about the physical world are subordinated both to the models and to the possible actions of experimental scientists, does this not lead to an apparently unsettling consequence? Since the final state of a cognitive system depends upon the path that led to it, the same final state could not have been attained if different paths had been followed. This would mean that an object, x, real or conceptual, would not be the same x, but another (x'), if the path (t) to its discovery had not been t, but $t'x'$, which could lead only to x'. Thus, one might conceive of the possibility that our knowledge about physics might have begun with biophysics instead of with inorganic matter, and hence would have led to an entirely different physics, subordinated from the start to the notion of field or of "dissipative structure" in the sense of Prirogine;[11] or that another mathematics might have come about if the Greeks had started from non-Euclidean or non-Archimedian postulates.

In contrast, if one contests these suppositions and attributes to the human mind a unique kind of physics and mathematics, this would lead us back to the hypothesis of predetermination of knowledge. Thus, it seems that we are presented with an unavoidable dilemma: either a heterogeneous diversity of possible paths and results or else the preformation of knowledge.

However, one might refuse to accept this dilemma. The different results obtained by these different paths will sooner or later be subject to coordination by means of transformations of some degree of complexity relating x to x'. Such coordinating transformations, however, cannot have been preformed; they are possible only after the fact. We are confident in the intervention of such coordinations, because nature (or the world) exists; no matter how unstable it may be (after all, it is composed of a continual evolution of beings and of mind) it is not contradictory for all that. Now, within this world, there exist two kinds of junctions that allow interactions between the mind and physical reality: a terminal junction, which allows the subjects to experience the external world, and an initial junction, which unites the mind to the body and the body to the structures of the physical world, with which it is in constant interaction—in almost a circular fashion. This implies that even though mental constructions travel far beyond the limits of the phenomena (even "infinitely" far), they may still be in concord with the latter so that if different paths of research have led to apparently incompatible results, there is some hope that the coordination will be possible, given the invention of new cognitive tools.[12] A good example is that of the wave and the particle in microphysics: initially, they were thought to be incompatible, while later, they were seen as associated. Today it is still a mystery how they are related to each other, but physicists are confident that sooner or later there will be an explanation. Now, for two reasons such coordinations cannot be thought of as predetermined. One concerns the mind, the other external reality itself. In the first place, the terms to be coordinated are novel and variable so that their interrelations cannot be established until the terms themselves have been constructed. In the second place, nature itself, no matter how dynamic and rich in transformations it may be, evolves also by successively opening up new variations as a function of modifications that also occur in real time. To say that all conceivable combinations are pre-formed in an initial set of "all" possible ones is therefore devoid of any sense, all the more so as "all" is simply a union set that must itself be thought of as "possible," referring both to this union class and to its subclasses.

VIII. THE NOTION OF FACT

The question concerning the relations between facts and interpretations will come up repeatedly in this book, which will treat, for the most part, the epistemology of physics. It may thus be useful to specify here the problems raised by the notion of "fact" as such.

Something observable, no matter how elementary, presupposes much more than a perceptual reading, since perception as such is itself subordinated to action schemes, which have logic in the interplay of relationships getting established (embeddings, etc) and which constitute the framework for all observation. Therefore, an observable is from the beginning a product of the union between a content given by the object and a form required by the subject as an instrument necessary in any reading of facts. Now, if this is true of simple registering of facts, it is obvious that the part played by the subject's constructions becomes increasingly important as one goes on to the different levels of interpretation. Let us recall that in fact, several such levels exist up to the terminal ones, which are characterized by the search for causal explanation.

We may, thus, consider a "fact"—which may be a property, an action, or any kind of event—as an observable once it has received an "interpretation"—that is, a meaning relative to a larger context. In contrast, the sense of a simple observable (any assimilation bestows a certain meaning) may remain entirely local in space and even in time. A fact, then, is always the product of a composition between a part furnished by the objects and another constructed by the subject. The part played by the subject can be important enough to lead to the deformation, even the repression, of the observable so that the fact is distorted by interpretation. For instance, children given the task of attempting to fasten a pebble to a string (a kind of a slingshot) to hit a box, while the string performs a rotation,[13] learn very quickly to release the pebble tangentially (i.e., in the 3 o'clock or 9 o'clock positions of the circle). But, when they are questioned about what they have done, they believe they have released the projectile in front of the box (in the 12 o'clock position). We are thus dealing here with a distortion of an observable caused by a false interpretation, according to which a moving object can reach its target only when propelled perpendicularly to the target.

Many other examples could be cited from developmental psychology, and everyone will agree that an analogous attitude can be found in the history of science, when a researcher, before accepting a fact that runs counter to a theory tries by all means to diminish its importance: even Planck, after discovering his first example of quanta, tried for several years to localize what he believed to be an error, before resolving to accept the fact. If the subject's interpretation is this evident even where it is erroneous, it is no less necessary to the process of constituting a fact when it is correct (and this at any level of cognition): the fall of an apple is just as much a fact for a farmer, who sees it as the normal consequence of the ripening of a fruit, as it was for Newton, who is supposed to have recognized such a fall as an example of gravitational attraction.

But even though the interpretation characteristic of the constitution of a fact already shows its close association with a system of concepts or sensorimotor schemes, this type of interpretation, while being superior to simple assimilation that registers only an observable, is nevertheless the least complex of a hierarchical series leading to the object, to legality, and finally to causal interpretation—i.e. to explanation. Observables can be registered either by means of a single scheme or by one that becomes differentiated by accommodation if necessary. We shall say that "facts" exist from the moment that a system of schemes becomes necessary: to say that that apple fell, that such and such mountain is covered with forests, or that the sun set at such and such a time in a particular place on a certain date is to say that all these are facts presupposing the coordination of concepts. But, even though there is progress compared to the elementary observables, this is still quite some distance from a deductive theory, be it causal or only legal.

Before getting to that point, let us note again that the "object" constitutes itself in interaction with the facts, hence at the same level, but with a duality of meaning. The formation of the object presupposes a coordination of facts—as is evident in the sensorimotor beginnings of object permanence, where positions or sequences that make possible the search for the object behind a screen intervene. But then, in turn, a fact, being associated with a system of assimilations, does not acquire its wider meaning, which distinguishes it from an observable, unless it expresses the properties or the actions of objects. We may, thus, consider the object as a synthesis of facts attached to a single substrate and the facts as expressions of relations or interactions between objects or within a single object (but in the latter case, with possible correspondences to other objects).

IX. LEGALITY AND CAUSALITY

Given this, further progress in the interpretations, which accompany the constitution of facts, is naturally due to a generalization leading to the establishment of "general facts." Now, a general fact is nothing more or less than a law. This is true even at the most elementary levels, where its logicalization remains qualitative, as when an infant discovers that any suspended object may be swung. But even a highly mathematized physical law remains at the level of a general fact, as long as its generality remains inductive in kind and does not result from necessary connections. In that case, the mathematization consists in nothing more than a set of formulas "applied" by the physicist to a content provided by experience. Even where the application of such a formula is to some extent dictated by the nature of the content, the latter is not yet equipped with necessary characteristics. This remains so provided that content is not inserted in a system of causal relations that transcend the frontiers of the facts and of laws, since such causal relations, in essence, take a fact as both a necessary consequence of other facts and as a source of new transformations.

The interpretation involved in the construction of a causal model is, thus, of a higher level than that of interpretations that are constitutive of facts and of laws. Both may, however, make their first appearance at the same point in their development. What causal interpretation contributes as new is then precisely the necessity "attributed" to sequences with respect to their content. This intrinsic necessity is very different from an obligation to apply a particular mathematical formula to a given content, in which case the latter is registered, but not yet explained. Let us recall that the necessity of sequences is something different from their generality and that when people speak of "causal laws" to designate sequences that are only regular (i.e. already general), this is, in fact, a misnomer: causality not only expresses the "fact" that *B* "always" follows *A*, but also stipulates the mode of transformation leading from *A* to *B*—for example, the transfer of mv and $1/2\ mv^2$ of one ball to the next in propulsion.

It, thus, seems necessary to specify the differences between a law and a cause. To do this, we must analyze the nature of cases intermediate between legal inferences and causal deductions. The establishment of general facts or laws can, in fact, give rise to two distinct kinds of inferences. First, there are the inductive inferences, which allow the physicist to establish correspondences between his measures and to derive from these general relationships in the form of functional equations. These constitute, then, the forms "applied" to the content, which determines their choice. This content, however, is at this level only registered and not yet necessary in itself. Second, once the laws are established in this way, they can then be coordinated, by means of new inferences, into a deductive system, in which the particular laws are inserted within the more general ones. In this case, the forms are interrelated by necessary connections, while still in the state of applied forms simply determined by their contents. As for the contents themselves, once their diverse degrees of generality have been stated (but only stated) the constitution of inclusion relations is permitted, but without the furnishing of any reasons—that is, explanations.

In contrast, causality begins when the transformations that have been described up to then as "facts" can be coordinated in structural systems whose laws of composition subordinate the system of categories to operational transformations which are conjoined by necessary relations forming closed systems, where each has its own logic: such as the structure of the group. In this case, there is a change from inclusive to constructive relations, or to put it more precisely to a classification in terms of algebraic models which, so constituted, become explanatory. But then the relations between form and content become modified. Now it is the content itself that becomes the locus and the source of necessary connections, since the objects, relations, and transformations of reality (thus, of the content) are no longer interpreted simply by possible translations into logicomathematical forms applied by the subject. They now become the operators of the algebraic structure adopted by reality. Such are the higher forms of the series of interpretations, which begins with the "act," and which we can now characterize by saying that in this case the operations involved are no longer simply "applied" to the objects, but rather are "attributed" to them as inherent in the algebraic structure of the real system. Certainly, the subject knows this sytem only by means of the models she constructs, among them the laws established so that the operations now attributed can be identical in part with those that had the status of applications only. What is new, however—and properly characteristic of causality—is that these operations are now integrated in structures characterized by the internal necessities of their compositions.

Let us further specify that causal explanation, like legality, can appear in various degrees of successive approximations: when the transformation that explains *how A* becomes *B* is discovered, the question of *why* this is so presents itself. Whenever progress is made in explaining "how" something happens, it is followed by an explanation of "why" it happens that way, and so on. Thus, we find intermediate levels of structuration, which some authors accept as being explanatory while others see them as only somewhat more advanced cases of legality. A good example is that of the first meteorological models constructed by Von Neumann and J. Charney, using the first computer. Some specialists considered it as nothing more than a system of equations permitting predictions, and so not going beyond the level of legality. Others have seen the transformations involved as explanatory, because the variables referred to the entire field of physics (particularly to hydrodynamics and thermodynamics). In our language, this means that the former group of specialists focusing on form, have seen only "applied" operations, while the latter group, considering the implication of content as sufficient, have recognized the operations as "attributed" to the real processes themselves. This opinion has since been generally accepted.

X. DISCOVERY AND JUSTIFICATION

Our viewpoint will be in agreement with one of the fundamental theses of genetic epistemology: the constructivist nature of notions like validity, necessity, justification, and knowledge.

Such a thesis is in opposition to those defended by authors belonging to the positivist and neopositivist schools of thought, who had a decisive influence on the philosophy of science during the first half of this century, particularly in the Anglo-Saxon countries. Their position was given its most characteristic expression in Reichenbach's book, *Experience and Prediction.*[14]

Reichenbach introduced a radical distinction between what he calls the "context of discovery" and the "context of justification."

The former refers to the processes involved in scientific discoveries, to the way scientists conceive of a new concept, construct new theories, or find an explanation for an as yet unexplained phenomenon. The latter refers only to the mode of validation of a concept or scientific theory—in other words, its rational justification and legitimation as part of the accepted system of knowledge.

The study of the context of scientific discovery belongs to psychology and to history. According to Reichenbach, it is foreign to the philosophy of science, which takes account only of the context of justification. One consequence follows from this distinction: in order to validate knowledge, one has to purge it of any connotation it might have acquired in the process of discovery. Justification thus requires a complete re-construction of the knowledge in question, directed at demonstrating its coherence (from the point of view of its internal logic) and its confirmability (from the point of view of its correspondence to reality).

The goal of the philosophy of science would then be the rational reconstruction of scientific thought.

One of the immediate objections to this kind of position is the failure of theory, in any of its formulations, to provide foundation for its own affirmations or method to guarantee its validity. But the main difficulties of accepting this position lie elsewhere.

Any theory which introduces a dichotomy between constructive processes and methods of validation in the elaboration of scientific theories makes reference, explicitly or implicitly, to a knowing subject, attributing to it a particular role throughout the process. This, incidentally, is true of all epistemology; what is not less clear is the role of the subject in the process of acquiring knowledge.

If one makes reference to a "natural" subject, it is necessary to verify if what one says about this subject corresponds to reality (it suffices to recall the criticism frequently addressed to the empiricist theories of knowledge: the role empiricists attribute to the subject in the acquisition of knowledge has never been verified empirically).

If, on the other hand, one refers to an "ideal" subject whose norms are conceptualized and formulated in accordance with a different type of philosophical speculation, it becomes necessary to show what are the criteria of validity used to accept such epistemological conclusions, what the reasons are for substituting an ideal subject for a "natural" one, and what the relations are between the two.

It is interesting to note that the comparative analysis of epistemological theories reveals a very great variety of characteristics proper to the epistemic subject. Unfortunately, these characteristics often contradict one another.

The idea of two "contexts," proposed by Reichenbach, has found many opponents among contemporary philosophers of science. Kuhn is one of the most representative of these; he sees in the history of the scientific theories an essential element to account for their acceptability and their justification.

In chapter 9 we shall refer to Kuhn's position in greater detail. For the moment, it may suffice to indicate that his position is, in a sense, open to the same criticisms. In fact, Kuhn believes that the normative foundations used by the "builders" of the different fields of knowledge are to be found in their spontaneous cognitive development. But then one has to make a decision as to whether or not the specialists in the various scientific disciplines are truly representative of the "natural" subject. If so, they would simply extend the mechanisms and rational norms of any natural subject; if not, scientists would be in a separate class, different from that of the "natural" subjects; their cognitive activities would follow epistemological rules different in kind from those used in nonscientific thinking.

Those who believe that scientists use and apply precise rules in each phase of their work (discovery, invention, verification, explanation), and in everything they effectively accomplish, cannot easily provide precise indications as to what those rules are. Scientists are only partly conscious of what they do. Aside from those who have specifically studied the theoretical foundations of their discipline, it appears that even the most original scientists may entertain incomplete, even false, notions as to the nature of mental structures. A good many physicists, orthodox positivists, have made their discoveries by using procedures that run counter to basic positivist tenets. All through history, scientists have used cognitive structures without ever becoming conscious of them. A classic example is Aristotle, who used the logic of relations, but failed to incorporate it in the construction of his own work on logic.

Thus, there is a long distance to cover between the spontaneous, unconscious use of structures, and their becoming conscious. The first (and often considerable) difficulty, often encountered even by scientists, is in determining the rules effectively used by the subject. One has to perceive the operations as they are used in actions, even though only fragments of them may be conscious and then in often deformed, badly registered, or incomplete form.

The consideration of the evolution of scientific theories in a framework such as that specified by Kuhn must confront a further difficulty: any knowledge, no matter how novel, is never a first, totally independent of previous knowledge. It is only a reorganization, adjustment, correction, or addition with respect to existing knowledge. Even experimental data unknown up to a certain time must be integrated with existing knowledge. But this does not happen by itself; it takes an effort of assimilation and accommodation, which determines the internal coherence of the subject.

Cognitive structures, inasmuch as they are an organization of knowledge, are essentially comparable to organisms whose present state is determined not only by their present environment but also by their ontogenetic and phylogenetic history. This does not rule out the normative character that such structures may have for the subject. But it is necessary to specify that, in the case of cognitive processes, there is another determining factor: cultural transmission. In other words, knowledge is never a state but a process, which is influenced by preceding developmental stages. Therefore, an historicocritical analysis becomes crucial.

We shall claim that the only factors that are truly universal in any kind of cognitive development—in the history of science as well as in psychogenesis—are functional in kind rather than structural. They have to do with the assimilation of novel elements into preceding structures and the accommodation of the latter to new, effective acquisitions.

It still needs to be stressed that the content and the "organs" of structuration of these universal functions change continually so that they become integrated parts of historical evolution. In other words, the continual transformation of knowledge proceeds by reorganizations and re-equilibrations, step-by-step, without preformation. The intrinsic necessity of the structures constitutes the product of successive acquisitions.

If our position is correct, it follows that scientific knowledge is not a new category—one fundamentally different and heterogeneous with respect to pre-scientific thinking and with respect to the methods inherent in the instrumental acts characteristic of practical intelligence.

The norms of science represent an extension of the norms of prior thought and action, but they incorporate two new requirements: internal coherence (or the total system) and experimental verification (for nondeductive sciences).

XI. THE PURPOSE OF THE BOOK

In one particular case, that of the evolution of physics from Aristotle until just before Newton, we have been able to establish a correspondence—indeed a very direct one—between the four historical periods (the two Aristotelian driving forces, the recourse to a single driving force, the discovery of the impetus, and that of acceleration) and the four stages in psychological development. In particular, we observe a striking construction and generalization, at about seven or eight years of age, of the idea of élan, in surprising analogy with Buridan's concepts. In this case, the parallelism in the evolution of concepts in history and in psychological development concerns the content of the successive forms of the concept. This is not surprising, since we are dealing here, in a way, with pre-scientific concepts. But it would evidently be absurd to try to generalize this type of parallelism of contents in the case of scientific theories proper, such as those which emerged between Newton's mechanics and Einstein's relativity. Thus, in the latter case, our main research efforts are directed, not at the contents of concepts, but at the common instruments and mechanisms of their construction.

What, then, is meant by "common mechanisms"? There are a

number of different kinds. The most general kind of mechanism relates, of course, to the nature of reasoning. At all levels of psychogenesis, and in the history of science, reasoning involves "reflective abstractions" (see above) as well as empirical ones (in physics, the two forms alternate continually, and in mathematics, only the "reflecting" form is used), or constructive generalizations as well as extensional ones.

A second general characteristic is that there is no cognitive elaboration in which subjects appeal to pure experience, since, as we have seen, observables are always interpreted, and a "fact" necessarily implies an interaction between the subject and the objects in question. It thus follows that any domain of knowledge, be it accurate or erroneous, includes an inferential aspect. In each sector of cognition, even in simply "taxonomic," i.e. in descriptive zoology and biology, the subject's contribution is indisputable, since the very form of the classifications implicates the "grouping" structure, or semi-lattice. Now, "groupings" are constructed by children as young as seven or eight years of age, but, of course, with a restricted number of elements, whereas in systematic biology the number of elements is infinite.

From these common instruments and mechanisms there springs a third variety, which is already operative in sensorimotor constructions and up to the highest forms of scientific thinking: this is a dual process of differentiation and integration characterizing every cognitive progress, with the two aspects eventually becoming conjoint.

Fourth, at all levels of knowledge, from the practical "know-how" of the sensorimotor stage to the most advanced theories, there is a search for "reasons" (for failures as for successes). In each case, finding a reason means relating the results obtained to a "structure" or coordinated schemata. For example, at about 9 to 12 months, infants discover that objects that seemingly disappear from view are nevertheless permanent.[15] They do so by relating the "disappearance" to a "group" (in fact, in the sense included in Poincaré's definition, a practical grouping) of displacements.

Fifth, it is remarkable to find, as we shall continually, that the advances made in the course of the history of scientific thought from one period to the next, do not, except in rare instances, follow each other in random fashion, but can be seriated, as in psychogenesis, in the form of sequential "stages." We shall try to describe, for each domain, the most important of these stages. For example, in the case of that fundamental period in the evolution of algebra which begins with the "groups" of Galois, we witness a series of constructions that is not at all random; each is made possible by the preceding ones and each in turn prepares those which follow. In geometry, it is no accident that the non-Euclidean geometries appeared a good deal later than the Euclidean ones and that coordinate systems were constructed much later than the analysis of geometrical forms. One of the essential aspects of our present research is to try to identify such laws of progression. But it needs to be made clear from the outset how we define the goal we wish to attain by comparing such progressions to those observed in psychogenesis: this goal is not to set up correspondences between historical and psychogenetic sequences in terms of content, but rather to show that the mechanisms mediating transitions from one historical period to the next are analogous to those mediating the transition from one psychogenetic stage to the next.

Now, the "transitional mechanisms," the main topic of the present volume, exhibit at least two common characteristics between the history of science and psychological development: one of these we have already treated elsewhere, but the second seems to us to be new. The first of these mechanisms consists in a very general process characterizing all cognitive progress; this is the fact that in each progression what gets surpassed is always integrated with the new (transcending) structure (which—even in biology—is far from being the case in domains other than cognitive development). The second transitional mechanism is one we have never studied before, but it will become central to the present volume. It seems to us likewise of a very general nature: it is a mechanism that leads from intra-object (object analysis) to inter-object (analyzing relations or transformations) to trans-object (building of structures) levels of analysis. That this dialectical triad can be found in all domains and at all levels of development seems to us to constitute the principal result of our comparative efforts.

In fact, the generality of this triplet, *intra, inter,* and *trans,* and its occurrence at all sublevels as well as within global sequences undoubtedly constitutes the best of the arguments in favor of a constructivist epistemology.[16] Empiricism might at best account for the transition from the intra to the inter level, since in this case, relations have to be substituted for initial predicates, and relations may be suggested by a simple reading of facts. But the transition from the inter to the trans object level involves going beyond the preceding stage, which requires necessary constructions and whatever is associated with these. As for the aprioristic position, it may be able to interpret the trans-object activities, which would be seen as preformed, but it would be incapable of explaining why the transitions in the sense of transformations should have to be prepared by intra and inter-object analyses. The unavoidable intra, inter, trans sequence clearly demonstrates the constructivist, dialectical nature of cognitive activities. We believe this demonstration has far-reaching consequences.

I

From Aristotle to the Mechanics of the Impetus

The main objective of this historical chapter is somewhat different from the later ones. In the later chapters, the very content of the themes whose history we are going to trace corresponds, with a few exceptions, to levels of abstraction which go far beyond those studied in developmental psychology. Our research will, for those cases, be directed toward the identification of the mechanisms operating at each transition from one stage to the next in the evolution of concepts and theories. This will be done for each of the fields of science chosen to illustrate and to "verify" the fundamental hypotheses that guide our overall conception. We shall, in each case, take up a particular science—i.e., one that is already constituted, even though not necessarily "completed." Our aim in the historical analysis will be to study its successes, its failures, its crises and their resolutions. Particular importance will be given to the mechanisms which make these resolutions possible.

In this chapter, however, we shall treat a topic which, up to the end of the Middle Ages, showed a development quite similar to the pre-scientific thinking of the child or the adolescent when confronted with such phenomena. This will not only enable us to establish relationships between mechanisms but also to examine the content of the concepts used.

The reason that such direct correspondence between contents is possible in one case, but not in others, represents in itself an epistemological problem, of which we shall here only indicate the characteristics and the scope. In the conclusions to this volume, we shall present an explanation for this problem.

We shall begin with Greek science. We do so not to imply that the Greeks were the only students of science, but simply because it is possible to establish a historical continuity, with a sufficient number of supporting documents, between them and contemporary science. Nevertheless, the historical process of the problem studied in this chapter is very different from the one of the following chapter, which deals with geometry.

In the case of geometry, Greek science already presents, with Euclid, Archimedes and Apollonius, a diversity of content and a structure which remained the same over the centuries. There were, to be sure, certain modifications in the nineteenth century, but they became integrated within a wider and more complex overall conception. In contrast, nothing has remained valid to this day of Aristotle's "Science of motion." The long, complex evolution that led to Galilei, Huygens, and Newton contains, as we shall see, a few precursors of the ideas developed in the seventeenth century. The emergence of mechanics at that time demolishes completely the conceptions of "dynamics" that had been elaborated during the preceding centuries. Mechanics, as a science (in the modern sense of the word), did not occur before the seventeenth century. This development, extremely slow relative to that of mathematics, makes it possible, owing to an abundant documentation, to follow, step-by-step, this long "pre-scientific" evolution, over which one can trace the emergence and development of the concepts *in statu nascendi.* Nothing of the sort can be done for mathematics, except for a few isolated topics; there we would have to go back to periods where written material does not exist in sufficient quantity to make such analyses possible.

Chapters 1 and 2 are thus the only ones where we shall look for correspondences in contents between certain concepts as they appear in psychogenesis and in the history of science. But we shall also find here the beginnings of a deeper correspondence in the mechanisms operating in the two kinds of processes, the psychogenetic and the historical. This parallelism, as well as the "primitive" nature of these mechanisms, which have placed, during many centuries, severe restrictions on conceptual development, have gone unnoticed by historians of science, who have failed to appreciate their importance to epistemology. As far as we are concerned, we have been led to re-evaluate these facts in the light of an examination, which first presented itself in the field of psychogenesis. From there, we proceeded to a rereading of the history of science from a perspective that differs from that of most specialists.

The choice of the historical material we consider remains partly arbitrary, especially in this chapter. Neither do we claim to present an exhaustive study of the themes taken up. The importance we give to the ideas of certain authors, the brief overviews we give of the ideas of others, and the omission of still other authors generally considered "important" would be unacceptable in a historical account that purported to rigorously present the facts, and evaluate their merits and significance. Our aim, again, is entirely different—and so are the criteria for our selections. Our criteria are, in particular, clarity of their presentation, how well they represent a period or a school of thought, and the lasting value of their ideas. Thus, among the Greeks, we have concentrated only on Aristotle's doctrine for two reasons. It is Aristotle who expresses with the greatest precision the conceptions on which we base our proposed historical-psychogenetic comparisons. In addition, those doctrines have provided the essential "system of reference" in the course of the later development of mechanics all the way to the beginnings of modern science.

I. THE ARISTOTELIAN DOCTRINE OF MOTION

1. Introduction

A valid physical theory rests generally on three kinds of conditions: (1) a methodology or use of methods designed to analyze the facts and verify hypotheses; (2) a set of epistemic positions (not necessarily in the form of a thematic epistemology) furnishing a specification of the general concepts used, in strict correspondence with the experimental data collected but transcending them to a lesser or greater extent at the level of intensional definition; (3) the construction of a coherent system in which the analyzed facts and the concepts used are related and logically integrated to a sufficient degree.

Now, in the case of Aristotle, the methodology is reduced to a certain number of direct, rather simple observations, limited by the process we shall call "pseudo-necessity." For example, the only movements he recognizes are rectilinear or circular, hence his absurd conclusions concerning the paths of projectiles. His epistemic positions are thus impaired from the outset, because of lack of experimental data. In contrast, the facts (rightly or wrongly considered such) and the concepts used to express them are related within a system of an impeccable logic. This explains its success over the centuries; not until Newton is it possible to find a system as coherent. The particular case of a theory of mechanics of which nothing has survived, but which presented itself within a system of such deductive force that the theory was assured a long-lasting success, certainly is worth examining in detail. This is all the more so as Aristotelian physics, lacking an experimental method and because of its "pseudo-necessary" submission to highly underdifferentiated observables, offers many possible parallels to the process of psychogenesis.

2. The Theory

In Aristotle's conception, motion is understood in a very wide sense: "Hence there are as many types of motion or change as there are meanings of the word "is" (*Physica,* III [1] 201 a).[1]

He arrives at this conclusion after having affirmed, contrary to Plato's doctrines (*Parmenides,* 138 B, 162 E; Sophist, 248 E), that "there is no such thing as motion over and above the things," which thesis, in turn, is justified in the following manner:

> It is always with respect to substance or to quantity or to quality or to place that what changes changes. But it is impossible, as we assert, to find anything *common* to these which is neither "this" nor *quantum* nor *quale* nor any of the other predicates. Hence neither will motion and change have reference to something over and above the things mentioned, for there is nothing over and above them. (*Physica* III [1] 200 B–201 A) (TN)

In essence, according to Aristotle, all change is a kind of motion. Each of the "modes of being" in the preceding citation "belongs to all its subjects in either of two ways: namely (1) substance—one in positive form, the other privation; (2) in quality, white and black; (3) in quantity, complete and incomplete; (4) in respect of locomotion, upwards and downwards or light and heavy." (*Physica* III [1] 201 A, 3–7).

In short, for each "mode of being" there are polar opposites through which it gets realized. Movement produces only a changeover from one to the other element in each pair.

This characteristic of Aristotle's doctrine reveals a very surprising lack of differentiation between the organic and the inorganic levels, between what is purely mechanical and what pertains to the physical or physicochemical domains, and even between the biological and the cognitive spheres. Even though each "mode of being" has its own "type of motion," and even though the various forms of motion are not comparable (*Physica,* VII [4], 248 A), Aristotle's analysis of the characteristics proper to purely mechanical motion—which he calls "transfer"—refers back to the other types of motion, applying to mechanical motion the observations and arguments that he derives from the other kinds of motion. Let us look at some examples. This is the opening passage of Book V of *Physica:*

Everything which changes does so in one of three senses. It may change (1) accidentally, as for instance when we say that something musical walks, that which walks being something in which aptitude for music is an accident. Again (2) a thing is said without qualification to change because something belonging to it changes, i.e. in statements which refer to part of the thing in question: thus the body is restored to health because the eye or the chest, that is to say a part of the whole body, is restored to health. And above all there is (3) the case of a thing which is in motion neither accidentally nor in respect of something else belonging to it, but in virtue of being itself directly in motion. Here we have a thing which is essentially movable: and that which is so is a different thing according to the particular variety of motion: for instance it may be a thing capable of alteration: and within the sphere of alteration it is again a different thing according as it is capable of being restored to health or capable of being heated. And there are the same distinctions in the case of the mover: (1) one thing causes motion accidentally, (2) another partially (because something belonging to it causes motion), (3) another of itself directly, as, for instance, the physician heals, the hand strikes. (*Physica* V [1] 224 A)

Further on, after having drawn a distinction between the initial and the final term of a motion and having affirmed that "it is the goal rather than the starting point of motion that gives its name to a particular process of change," Aristotle adds:

> Here also the same distinctions are to be observed: a goal of motion may be so accidentally, or partially and with reference to something other than itself, or directly and with no reference to anything else: for instance, a thing which is becoming white changes accidentally to an object of thought, the colour being only accidentally the object of thought; it changes to colour, because white is a part of colour; or to Europe, because Athens is a part of Europe; but it changes essentially to white colour. (*Physica* V [1], 224 B, 16–23)

In other places, Aristotle analyzes, as examples of change that make it possible to comprehend the nature and characteristics of motion, the act of learning or teaching. This does not mean that Aristotle attributes to inanimate things the same cause of movement as he does to living things. In fact, with regard to things in which "motion is natural," he states: "It is impossible to say that their motion is derived from themselves: this is a characteristic of life and peculiar to living things" (*Physica* VIII [4], 255 A). Nevertheless, he also attributes an "internal motor" to these things, as we shall see.

Now, what is motion? To account for it Aristotle draws a distinction that has become famous:

> We may start by distinguishing (1) what exists in a state of fulfilment only, (2) what exists as potential, (3) what exists as potential and also in fulfilment—one being a "this," another "so much," a third "such," and similarly in each of the other modes of predictation of being. (*Physica* III [1], 200 B, 25–30)

This distinction between two forms of being, the potential and the actual, forms the basis of Aristotle's definition of motion and also of his solution to one of the paradoxes in the Eleatic school. Motion is explained not as the passage from nonexistence to existence, but as a change in the mode of existence: the transition from the potential to the actual. The following excerpt from Physica concerning this point is well known:

> We have now before us the distinctions in the various classes of being between what is fully real and what is potential. The fulfilment of what

exists potentially, in so far as it exists potentially, is motion—namely, of what is alterable qua alterable, alteration: of what can be decreased (there is no common name), increase and decrease: of what can come to be and can pass away, coming to be and passing away: of what can be carried along, locomotion. (*Physica* III [1], 201 A)

In Book VIII of Physica, Aristotle gives some interesting specifications on the subject, after recognizing that the expression "potentially" can be understood in more than one way:

One who is learning a science potentially knows it in a different sense from one who while already possessing the knowledge is not actually, exercising it. Wherever we have something capable of acting and something capable of being correspondingly acted on, in the event of any such pair being in contact what is potential becomes at times actual: e.g. the learner becomes from one potential something another potential something: for one who possesses knowledge of a science but is not actually exercising it knows the science potentially in a sense, though not in the same sense as he knew it potentially before he learnt it. And when he is in this condition, if something does not prevent him, he actively exercises his knowledge: otherwise he would be in the contradictory state of not knowing. In regard to natural bodies also the case is similar. Thus what is cold is potentially hot: then a change takes place and it is fire, and it burns, unless something prevents and hinders it. So, too, with heavy and light: light is generated from heavy, e.g. air from water (for water is the first thing that is potentially light), and air is actually light, and will at once realize its proper activity as such unless something prevents it. The activity of lightness consists in the light thing being in a certain situation, namely high up. (*Physica* VIII [4], 255 A–B)

Let us now examine in greater detail the most important characteristics of the Aristotelian doctrine of motion. Given our goal, we shall consider the following distinctions:

1. There are two kinds of motion: natural and compulsory or violent motion. Aristotle arrives at this distinction by a somewhat curious avenue. Violent motion, according to him, is immediately evident: the throwing of a stone is a sufficient demonstration. From this he infers that there must also be a natural motion, "since the complusory movement is contrary to nature, and movement contrary to nature is posterior to that according to nature" (*Physica* IV [8], 215 A). This latter principle, then, leads him to state that "if each of the natural bodies has not a natural movement, none of the other movements can exist" (ibid).

2. The second property of the Aristotelian doctrine of motion concerns the nature of material bodies. "Ici bas," in the "sublunar world," there are only four simple elements: earth, water, air, and fire. All bodies in the sublunar world are made out of one of these simple elements or out of a mixture of these elements, in varying proportions. In addition, there is a natural order among these simple elements; they are ranked in the order in which we have just mentioned. For this reason, each simple body, when it is away from the place that belongs to it, is endowed with a natural movement which incites it to return to its proper place. Such a motion is rectilinear and moves either toward the center (downward), as in the case of earth and water, or outward (upwards), as in the case of air and of fire. It is because of this natural tendency that the earth and the water are "heavy," while the air and the fire are "light."

Although "fire . . . has no weight, nor does earth have any lightness" (*De Caelo,* IV, 4, 311 B), the situation concerning air and water is different:

> Neither of them is absolutely either light or heavy. Both are lighter than earth—for any portion of either rises to the surface of it—but heavier than fire, since a portion of either, whatever its quantity, sinks to the bottom of fire; compared together, however, the one has absolute weight, the other absolute lightness, since air in any quantity rises to the surface of water, while water in any quantity sinks to the bottom of air. (*De Caelo,* IV [4], 311 A)

The distinction established by Aristotle between water or air as "intermediate" elements and fire or earth as extreme elements is very surprising, and in some cases difficult to interpret. The most curious of these distinctions is probably the one referring to the reciprocal nature of the predicate "similar." In a fairly controversial passage, he says:

> adjacent elements to one another resemble each other: thus, water is similar to air and air to fire; and among the intermediate elements, the relation may be reversed, but not between the extremes. (*De Caelo,* IV [3], 310 A)

It is interesting to note that earth itself plays no part in this motion. The center of which Aristotle speaks, is the center of the universe and even though the center of the universe coincides with that of the earth, it is toward the former and not the latter that the heavy bodies move. A long passage is designed to explicate this fact (De Caelo, II, [14], 296 B), but the following citation is the one that is really conclusive:

Now, that which produces upward and downward movement is that which produces weight and lightness, and that which is moved is that which is potentially heavy or light, and the movement of each body to its own place is motion towards its own form. (It is best to interpret in this sense the common statement of the older writers that "like moves to like".) For the words are not in every sense true to fact. If one were to remove the earth to where the moon now is, the various fragments of earth would each move not towards it but to the place in which it now is. (*De Caelo* IV [3], 310 A–B)

To conclude, let us note that in *De Caelo* the simple elements are defined on the basis of each kind of motion:

Bodies are either simple or compounded of such; and by simple bodies I mean those which possess a principle of movement in their own nature, such as fire and earth with their kinds, and whatever is akin to them. (*De Caelo* I, [2] 268 B)

3. One of the most important characteristics of Aristotelian doctrine is the concept of a "motor" that can be identified with the cause of motion. Now, the distinction previously established between natural and compulsory (violent) movement leads to a further parallel distinction between two kinds of "motor":

Now it is impossible to move anything either from oneself to something else or from something else to oneself without being in contact with it: it is evident, therefore, that in all locomotion there is nothing intermediate between moved and movent." (*Physica* VII, [2], 244 A–B) ("Movent" is traditionally used in these contexts—TN).

As far as movement is concerned, "nature is a source of movement within the thing itself" (*De Caelo* III, [2], 301 B), which brings us back to the definition already given in *Physica* (II, [1] 192 B), where Aristotle indicates that nature is a source or a cause by virtue of which anything whatever is moved or remains at rest, by virtue of itself and not "by virtue of a concomitant attribute." In the latter text, Aristotle nevertheless emphasises that we should distinguish the nature of a thing from the attributes which belong to it by virtue of what it is, as in the case of fire. That "element" has the property of moving upward, which it does not possess "by nature," but which comes about "in accordance with nature." Natural motion has, thus, an intrinsic cause ("internal motor"). The "contact" between the motor and the mobile is evident in this case. Now, the "violent" movements proceed from an external cause ("external motor"), from a force which drives it to displacement in spite of its own nature:

But since "nature" means a source of movement within the thing itself, while a force is a source of movement in something other than it or in itself qua other, and since movement is always due either to nature or to constraint, movement which is natural, as downward movement is to a stone, will be merely accelerated by an external force, while an unnatural movement will be due to the force alone. (*De Caelo,* III [2] 301 B)

4. All motion needs an environment in which the mobile can move about. Only in this way was Aristotle able to explain why violent movement continued after the force that had given it its initial propulsion had ceased to act upon it. The difficulty of the problem is recognized in the *Physica* VIII, [10], 266 B).

We must consider that although the force that impels a projectile produces, according to Aristotle, a "movement," this movement corresponds, in cinematic terms, to the notion of *speed.* The great advance made by mechanics in the seventeenth century was to introduce explicitly the notion of inertia, and thus to relate force to acceleration rather than speed. Aristotle, to whom the very idea of inertia seemed absurd, as we shall see further on, thought that when the force has stopped acting, the movement should stop also. Now, since the movement continues, he is forced to look for another "motor" which maintains contact with the mobile, thus becoming the cause of its displacement. The following quotation condenses the well known explanation, which remained the accepted doctrine for many centuries:

"If everything that is in motion with the exception of things that move themselves is moved by something else, how is it that some things, e.g. things thrown, continue to be in motion when their movent is no longer in contact with them? If we say that the movent in such cases moves something else at the same time, that the thrower e.g. also moves the air, and that this in being moved is also a movent, then it would be no more possible for this second thing than for the original thing to be in motion when the original movent is not in contact with it or moving it: all the things moved would have to be in motion simultaneously and also to have ceased simultaneously to be in motion when the original movent ceases to move them even if, like the magnet, it makes that which it has moved capable of being a movent. Therefore, while we must accept this explanation to the extent of saying that the original movent gives the power of being a movent either to air or to water or to something else of the kind, naturally adapted for imparting and undergoing motion, we must say further that this thing does not cease simultaneously to impart motion and to undergo motion: it ceases to be in motion at the moment when its movent ceases to move it, but it still remains a movent, and so it causes something else consecutive with it to be in motion, and of this again the same may be said. The motion begins to cease when the motive force produced in one member of the consecutive series is at each stage less than that possessed by the preceding member, and it finally ceases when one member no longer causes the next member to be a movent but only causes it to be in motion. The motion of these last two—of the one as movent and of the other as moved—must cease simultaneously, and with this the whole motion ceases. Now the things in which this motion is produced are things that admit of being sometimes in motion and sometimes at rest, and the motion is not continuous but only appears so: for it is motion of things that are either successive or in contact, there being not one movent but a number of movents consecutive with one another: and so motion of this kind takes place in air and water. Some say that it is "mutual replacement." (*Physica* VIII [10] 266b–267 A)

Once the "mechanism" that assures the continuation of motion was introduced, the coherence of his own reasoning forces Aristotle—in the last part of the text quoted—to renounce its continuity: movement is only apparently continuous. The succession of the "substitutions" of air, which propels the mobile, presupposes a succession of "motors"; and the substitution of a motor for another, even if it is instantaneous, implies a cessation, similarly instantaneous, of the force, and consequently of the motion itself. It is thus not a single motion, but a "series of consecutive motions." The only continuous motion is that produced by the "immobile motor"; this is continuous motion because the motor always remains invariable so that its relation to what it moves also remains invariable and continuous. Here, as elsewhere in his work, Aristotle—logician above all—accepts without hesitation the logical consequences of the premises which he introduced into his reasoning.

Now, the air does not act only in the case of horizontal or oblique motions of projectiles. In the fragment already cited (*De Caelo,* III [2] 301 B), where he distinguishes between the effects of a force in the case of a natural movement and that of a violent one, Aristotle adds:

> In either case the air is as it were instrumental to the force. For air is both light and heavy, and thus qua light produces upward motion, being propelled and set in motion by the force, and qua heavy produces a downward motion. In either case the force transmits the movement to the body by first, as it were, impregnating the air. That is why a body moved by constraint continues to move when that which gave the impulse ceases to accompany it. (*De Caelo* III [2] 301 B)

5. Motion in empty space is impossible. To demonstrate this, Aristotle makes use of two arguments, one applicable to natural, the other to compulsory motion. With respect to the former, Aristotle asks "how there could be a natural movement where there are no differences." In fact, "in the void, what is high differs in no way from what is low, for as there is no difference in what is nothing, there is none in the void (for the void seems to be a nonexistent and a privation of being)." In contrast, natural movement includes differences, and the natural things contain differences by nature." From this, there follows the logical consequence expressed in the form of an alternative: "either then there is no locomotion or else then there is no void."

As for the compulsory motions, Aristotle takes the example of the projectile:

> Further, in point of fact things that are thrown move though that which gave them their impulse is not touching them, either by reason of mutual replacement, as some maintain, or because the air that has been pushed pushes them with a movement quicker than the natural locomotion of the projectile wherewith it moves to its proper place. But in a void none of these things can take place, nor can anything be moved save as that which is carried is moved. (*Physica* IV [8] 215 A)

On this point, he adds an argument, which he then refutes in a surprising manner:

> Further, no one could say why a thing once set in motion should stop anywhere; for why should it stop *here* rather than *here?* So that a thing will either be at rest or must be moved ad infinitum, unless something more powerful get in its way. (*Physica* IV [8] 215 A)

We are here obviously very close to the principle of inertia, which would have been formulated correctly if it were not for the expression "it must be moved ad infinitum," which is ambiguous: ad infinitum could be taken to mean *perpetually* (the object would never stop), but within the general context of Aristotle's thinking it must be interpreted as meaning that the object would acquire an *infinite speed.* However, Aristotle considers this idea absurd, and so repudiates it. From that he derives the conclusion that the void does not exist.

6. There are only two simple natural motions: the rectilinear and the circular. "The reason for this is that these two, the straight line and the circular line are the only simple magnitudes" (*De Caelo,* I [2]). As for the natural movements the circular movement "is the movement that turns around the center" (what is meant, to be sure, is the center of the universe); whereas the movement in a straight line is "the one which is directed upwards or downwards." But these two motions are not considered to be at the same level. The circular motion is perfect, while the straight line motion is not:

> this cannot be said of any straight line: —not of an infinite line; for, if it were perfect, it would have a limit and an end: not of any finite line; for in every case there is something beyond it, since any finite line can be extended. (*De Caelo* I [2] 269 A)

3. An Example of Aristotelian Reasoning

Aside from the contents of Aristotle's ideas about motion, it is interesting to analyze the type of reasoning he used to arrive at his conclusions.

To make this clearer, let us take a typical, representative example of Aristotelian reasoning. The title of chapter 2 of book I of the treatise *De Caelo* is the following: "Demonstration of the existence of a fifth element, endowed with circular motion." Instead of reproducing here the text in its entirety, it seems preferable to indicate explicitly the premises upon which he bases his deductions and to show how he arrived at his conclusions.

a. Premises:

(1)All motion "according to the location," and which we shall call translatory, is rectilinear, circular, or a mixture of the two.

(2) All simple translatory motion, either moves away from the center or else turns around the center.

(3) Simple motion is the movement of a simple body.

(4) The movements of simple bodies are simple, and the movements of combined bodies are mixed.

(5) For each simple body there is one and only one natural movement.

(6) Upward and downward motion are opposites.

(7) A single thing has only one opposite.

(8) The circle is perfect.

(9) The straight line is not perfect.

(10) The perfect is prior to the imperfect by nature.

b. Demonstration:

(1) Circular motion cannot be the natural movement of one or the other of the four sublunar elements [by (5)].

(2) It cannot be the natural movement of a combination of these elements, either [by (4)].

(3) It must be the movement of a simple body [by (3)].

(4) It must be a natural movement, otherwise there would result one or the other of two consequences, both false: (a) If the body whose movement is circular is part of fire or another element of the same kind, its movement would be the opposite of circular motion. But this is impossible because of (6) and (7). (b) If, on the other hand, the body advanced by a non-natural motion is something other than the elements, it should have some other motion that would be natural for it. Now, this is impossible, since if it were upward motion, the body would be fire or air, and if it is downward, it would be water or earth. And this has already been demonstrated to be impossible.

(5) It must be anterior to rectilinear motion [by (8), (9), and (10)].

Aristotle concludes: "On all these grounds, therefore, we may infer with confidence that there is something beyond the bodies that are about us on this earth, different and separate from them; and that the superior glory of its nature is proportionate to its distance from this world of ours." The long argument which precedes this is impeccable from the point of view of logical reasoning, and the conclusions impose themselves (as necessary), once the premises are accepted.

4. The Characteristics of Aristotle's Physics

Aristotle's physics does not take as a starting point the study of certain particular types of motion; instead it proceeds from certain general metaphysical principles. Aristotle does not analyze how bodies descend in free fall (Galileo would do so two thousand years later). He begins with a general observation: *the fact that bodies fall.* Then, he tries to *infer* how they fall, by means of rigorous reasoning based on metaphysical principles. The conclusions at which he arrives are completely implausible. The most elementary empirical observation would have been sufficient to invalidate it. Nevertheless, the fact that he arrived at these conclusions is less surprising, if we analyze carefully the internal logic and the epistemological foundations of the Aristotelian system.

One can get to the nature of things using this type of approach, i.e. immediate, general and qualitative experience. After this, one can infer the way bodies behave, which needs to be consistent with the nature of things as well as the principles, which are also very general, not verifiable, and imposed by the mind (for example, the principle that what is perfect must necessarily precede what is imperfect). This is why Aristotle does not really need to verify if heavy bodies fall faster than light ones. This is a consequence that follows from his principles and other general observations. He therefore has to accept it and so at no time does he feel the need to verify it by an experiment. This should not surprise us exceedingly: neither Galileo nor Huygens verified the consequences of principles of which they were firmly convinced.

Looking at it this way, one cannot consider tenable the position usually taken in the classical writings on the history of science and their explanations as to the nature of the differences between the mechanics of the seventeenth century as developed by Galileo,

Descartes, Huygens, and Newton and that of Greek antiquity and the medieval period. This classical position has been partly corrected in the course of the last 25 years, as we shall see below.

The majority of traditional writings on the topic suggest that the difference between Aristotelian and modern science resides in the use made of observation: while Galileo and his followers elaborated their theories on the basis of observations and experiment, the ancient and medieval worlds sought answers to the questions about nature in speculative reflection or meditation without even verifying their conclusions by experience. Such an account of the differences is only partly correct, and from an epistemological point of view it ignores the essential mechanisms of the slow and complex process that has led to the constitution of what today we call "scientific knowledge." We shall defend the thesis that the difference between ancient and modern science in no way lies in a willingness or unwillingness to resort to empirical observation, nor in the use of or abstinence from deductive methods. The explanation has to be sought elsewhere. But, before coming to that point, we will have to go somewhat further in the analysis of the historical facts themselves.

5. Criticisms of Aristotle's Doctrine

Aristotle's theories about the movement of projectiles were to be strongly refuted by Philopon, during the fifth century A.D., in his Commentaries on Aristotle's Physics.[4] Philopon considered any of the possible forms of antiperistasis as implausible—pure fantasy. In fact, the air is thought to carry out three different motions: it is supposed to move forward, pushed by the projectile; then it is supposed to turn around and go to the back of the projectile ("as if obeying an order"); finally, it has to change direction again and move forward, pushing the projectile. Philopon asks: how is it possible that the air does not get scattered, and how is it possible that it should hit precisely the back of the projectile? Furthermore, what is that force which imparts to the air, when it is first moving forward, the impulse that makes it change direction?

Philopon refutes Aristotle's second theory with an example, which can also be applied to the case of antiperistasis. If the air really causes the movement of the projectile once it has been launched, then why is it necessary for the stone to have been in contact with the hand or for the arrow to have been contact with the string of the bow? It would be sufficient to have a machine capable of setting the air in motion behind the stone or the arrow for them to start moving without any other contact except with the air. However, the truth is that "even if one sets in motion all the air behind the projectile with all the force possible, the projectile still could not move."

Philopon comes near to a much more modern conception when he says: "Whoever throws an object imparts to this object a certain action, a certain capacity to displace itself, which is incorporeal."[5] He also attacks the Aristotelian ideas about the void:

"Nothing prevents someone from throwing a stone or shooting an arrow, even if there is no environment other than the void. The environment impedes the movement of projectiles, which cannot advance without dividing it; yet they move within the environment. "Nothing, then, prevents an arrow, a stone, or any other object from being launched in the void; in fact, all is present, the movent, the mobile, and the space, which is to receive the projectile" (ibid).

These considerations, however, have no impact and do not prevent Aristotle's theory from being accepted. This fact is not trivial. When we read Aristotle's explanations and Philopon's refutation today, it may seem "obvious" that any "normal adult" would accept the latter's position. Aristotle's explanation of the movement of projectiles should definitely have been abandoned. But this has not happened. Simplicius (sixth century) attacks Philopon in his *Digressions Against Jean the Grammarian,* joined to his comments on Aristotle's *Physica,* which were translated into Latin in the thirteenth century. The Medieval Western Christian world would thus ignore Philopon's ideas, known only through the distorted version of Simplicius. However, in the Arab world, there is a tradition of thought which came to be influenced by Philopon's ideas and transmitted by Avicenne (980–1037), but there is no indication of a continuity with the school of the theory of "impetus," which flourished in Paris in the fourteenth century. It is within this school that Buridan was to recapture Philopon's arguments, without knowing that he had. Once again, the "obvious" had to wait several hundred years before becoming obvious.

II. MEDIEVAL MECHANICS

The next period in our historical survey is the Middle Ages; more precisely, the last centuries of that period. The "choice" of periods is determined, as we have noted, by reasons of practical limitation as much as by our goals.

1. The Persistence of Aristotle's System

Aristotle, Euclid, Archimedes, Heron, and Ptolemeus were translated into Latin during the twelfth century, at first from the Arabic versions. In the thirteenth century, the Western Christian world became acquainted with the majority of the Greek works that survived. In the fourteenth century translations of these works into the vernacular languages (French, Italian, Spanish) began to appear. Aristotle's system continued via the Arabic world and reappeared in western Europe in the twelfth and thirteenth centuries. It would remain influential until the seventeenth century, especially for the following reasons:

a. Aristotle provides the conceptual framework serving as a frame of reference for all reflection concerning science, and each new idea almost invariably had to be presented in the form of commentaries on the classical texts.

b. Aristotle indicates what kind of question one should ask about motion—that is, what the questions are which should be answered in any study of the movements of bodies.

c. Aristotle establishes the kind of "explanation" to look for, having introduced the idea of explaining nature in rational fashion by logical demonstration based on accepted premises (which, in themselves, could not be demonstrated, however).

Within this general frame of reference, the period we are concerned with is characterized by profound discussions concerning the scientific method. The theological substratum is always present, not only because those who practiced this kind of reflection were, with a very few exceptions, members of the Church, but, also because no free thought was allowed to express itself—that is, thought free from ecclesiatic censorship. But even under these conditions, there are cases—e.g., William of Ockham (1280–1347)—where the analysis of the sense of theoretical elements and of the very methods of scientific research were pushed to the limits of heresy. The scope and depth of these reflections were such that one can confidently affirm that the methodology of science up to the nineteenth century has not added a great deal. To illustrate how well-founded this affirmation is, we could give many examples, but it suffices to refer to the work of A. C. Crombie, *Robert Grosseteste and the Origins of Experimental Science* 1000–1700, where one can find surprising texts by Gorsseteste and his successors, Albert le Grand, Roger Bacon, Pierre de Maricourt, Wittelo, etc. concerning physics; in particular, on the role of inductive thought, the experimental character of that science, the relation between hypotheses and deduction, and the tendency toward mathematization.

From this point of view, the scholastic tradition represents a considerable advance with respect to Aristotle's methodology. It goes back to certain aspects of Plato's thinking, alternately contrasting or synthesising Plato's and Aristotle's positions. Still, this "going beyond" Aristotle's position brings about only a partial loss of his authority. His Physics remains the only coherent system for trying to explain the Universe and its phenomena in their entirety. But his writings are no longer uncontested, at least from 1277—the date when the Bishop of Canterbury condemned Aristotle's doctrines.

Actually, the Aristotelian system was offensive to the Church in only one aspect (the one Averroès had overemphasised): Aristotle's physics was too deterministic to leave sufficient freedom for divine intervention. It is thus the Averrian version of Aristotle's Physics which was condemned. Yet from that point on, the rest of the master's statements become open to doubt, even though they cannot resist the least confrontation with experience.

If one combines the emphasis on more thorough methodological analyses with the liberation from the master's authority, one might expect (after three centuries of reflection) that the science of motion would finally make a significant breakthrough. But this is not the case. We have already noted, for example, Aristotle's attribution of extraordinary properties to the movement of air, which Philopon had already considered absurd in the fifth century A.D. In the middle of the sixteenth century, we still find a man of Tartaglia's caliber affirming that if a cannon successively propels two missiles (under identical loading conditions, etc.), the second would go further than the first because it would "find the air divided, and consequently, easier to penetrate."

Before getting to the core of the epistemological problem, we have to take a closer look at the contributions made during this period in Medieval history to the *content* of the science of motion. We shall focus on three of these contributions, undoubtedly the most significant ones: The theory of the impetus; the modifications introduced to the Aristotelian law of the dynamics of motion; and the development of cinematics.

2. The Theory of the Impetus

Philopon's criticisms of the Aristotelian explanation of the motion of projectiles are taken up again by Franciscus de Marchia, and especially Jean Buridan and his followers, particularly Nicole Oresme (1320–1325/1382). The research carried out by A. Maier on the subject leaves no doubts as to the independence of Buridan's elaborations with respect to Marchia, as well as both Buridan and Marchia in relation to the Arabic successors of the school of Avicenne. Maier's conclusions are accepted today by the historians of Medieval Science (e.g., A-C. Crombie).

a. Buridan

Buridan gives a number of arguments against the Aristotelian theory of antiperistasis, all based on experience.

1. Consider a wheel (for example, in a mill), which keeps turning for a while after having been set in motion. Evidently, in this case one cannot speak of displaced air turning around and pushing again.

2. An arrow that has been sharpened at its rear end does not move more slowly, once it is released, than another unsharpened

arrow, which is contrary to the predictions of the theory ("probably the air which follows the arrow could not push a sharpened end, since it could easily be divided by the pointed rear").

3. A rowboat moving upstream on a river continues its motion for a certain time, even after the action of the oars has ceased. Contrary to the predictions, a man standing at the stern of the boat does not feel the air that "pushes" the boat but; actually, he feels the air from the front, which resists the movement.

4. Assume that the boat is carrying grain or wood and that a man is stationed behind the cargo. If the air had an impetus strong enough to move the boat forcibly, the man would be violently crushed between the cargo and the air pushing from the rear.

5. An athlete ready for a jump runs a certain distance to gather momentum, but once off the ground does not feel pushed by the air; on the contrary, the air is felt in front as resisting the athlete's motion.

Buridan was to take on to refute Aristotle's second theory as well, on similar grounds. The mill wheel and the boat serve once again as counterexamples. He also adds a third example, noting that Aristotle's theory "also predicts that one can throw a feather further than a stone and a lighter object further than a heavier one, size and shape being equal. Experience shows that this is false. The consequence is evident, since the air, once in motion, should support, carry, or propel a feather more easily than a heavier object."[7]

From these arguments, Buridan concludes that the movement impresses a certain *virtus motiva* (moving force) upon the stone or any other projectile—a certain impetus which acts in the same direction as the movent setting it in motion "either upwards or downwards, laterally or in circular motion."

The impetus Beridan speaks of has three important properties which distinguish it with respect to preceding theories:

a. The greater the speed of the driving force moving a body the more intense will be the impetus the body receives.

b. The larger the moved body, the stronger will be the impetus received by it.

c. The impetus is of a permanent nature (*res naturae permanetis*), and is not "spoiled" by the resistance offered by the surrounding medium.

The first two properties have induced some historians of science to take Buridan's impetus as an anticipation of the Newtonian concept of quantity of motion (the product of speed and mass). The third property, in turn, has been linked to a primitive notion of inertia. However, in the text cited above, Buridan refers to an impetus describing a "circular motion"; this idea represents an important obstacle in the development of the final conception of inertia. (Let us emphasize, nevertheless, that even Galileo did not have very clear ideas on this point. According to Galileo, the ideal case of inertial motion is one that is perfectly horizontal throughout its trajectory, that is, a circle concentric with the surface of the sea.)

In any case, in Buridan's system we find the idea, however diffuse, of energy associated with motion. We have to take into consideration the fact that concepts such as the quantity of movement and of kinetic energy developed very slowly in the course of history and that all kinds of hesitations concerning a precise definition remained until the seventeenth century (the proof is the debate between Liebniz and the Cartesians).

From these criticisms of his predecessors, Buridan develops the premises which permit him to elaborate his own conception of the fall of bodies:

a. The natural gravity of the stone remains always the same, before, during, and after the movement. Consequently, the stone weighs the same after the movement as before.

b. The resistance of the medium (i.e., the air) remains the same or nearly so during the entire fall. (Buridan adds not only that it seems false to him that the air near the ground should be less resistant than the air at the higher levels, but also that the latter should actually be less resistant than the former, being more delicate).

c. If the body in motion is the same, the "total mover" is the same, and the resistance also the same or similar, the movement will maintain the same speed, since the ratio of the movent to the moved to the resistance remains constant.

d. But it is a corroborated fact that the speed increases continually during the fall of a heavy body.

e. Consequently, "it is necessary to conclude that another force contributes to motion, aside from natural gravity, a force which moves from the start and which remains always the same." "Hence," Buridan continues, "we must suppose that a heavy body not only gains movement in itself from its principal motor—that is, its gravity—but that it also gains a certain impetus in itself, with that same movement. This impetus has the capacity to move the heavy body together with the natural, permanent gravity. And, since this impetus is confounded with the movement, it follows that the faster the movement, the more this impetus augments and gains in strength. Consequently, a heavy body is at first moved only by its natural gravity; so it moves slowly at the start. Then it is moved simultaneously by the same gravity and by the impetus gained; so, it moves faster. And since the movement gets faster, the impetus increases also and gets stronger. Thus, heavy bodies are moved by their natural gravity, and simultaneously by a bigger impetus. Hence, again it will gather speed. In this way, it will always continually accelerate until the end. And, to the extent that the impetus gets confounded with the movement, so it will diminish and get weaker when the movement subsides."

There are two things of note in this passage. First of all, it appears that Buridan remains faithful to the Greek idea that force produces speed (rather than acceleration). Hence, an increase in speed, such as that occurring during free fall, must necessarily be explained by an increase in force. In addition, the impetus here becomes ambiguous and the text itself becomes difficult to interpret. In fact, the impetus is generated, together with the movement, by the initial motor (gravity), and once generated, it *produces* more movement (greater speed); but it also appears that it is the movement itself which can generate more impetus ("and since the movement becomes faster, the impetus, too, increases and becomes stronger").

b. Oresme

Nicole Oresme is undoubtedly the most important adherent of Buridan's school. We have noted above that according to Aristotle (*De Caelo*) a free falling body increases in speed and therefore its weight. In commenting upon that assertion, Oresme clearly exposes this conception of impetus:

But here it has first to be noted that the speed of the movement of a heavy body does not always increase in the fall, since if it is made of denser material or material harder to separate at the bottom than on top, that could be so strong that the body could fall more slowly in the end than in the beginning so that the (total) speed would always be the same. Similarly, when he says that the weight increases proportionately to the speed, this is not to be understood as if weight as a natural quality would incline near the bottom. For if a one-pound stone fell from a very elevated place and its motion were far faster in the end than in the beginning, still the stone would not change its natural weight. *Rather one has to interpret this weight that increases in falling as an accidental property, caused by the greater strength in the increase of speed, as I declared before concerning Physics. This property may be called impetuosity.* It is not strictly speaking weight at all, since if there was a passage from here to the center of the earth and beyond and if a heavy object descended by this passage or hole, until it got to the center, it would pass through to the other side and rise up again by means of this accidental, acquired property. Then it would descend again and this would go on several times in the manner of a heavy object suspended by a string from a trevire. So, it is not really weight, since it makes objects rise as well. And this quality is in all motion, both natural and induced, every time there is an increase in speed, force or movement in the sky. This quality is the cause of the motion of objects that are launched once they have left the hand or the projection device, as I have shown before in the chapter on Physics.[8]

The parts we have italicized leave no doubt as to Oresme's conception of the impetus as a product of the acceleration of bodies. But then, the impetus itself produces additional acceleration. The characteristization of the impetus as simultaneously cause and effect serves Oresme as the point of departure for his demonstration of the traditional theory of acceleration in projectiles at the beginning of their trajectory.

III. EPISTEMOLOGICAL REFLECTIONS

Joseph T. Clark, S. J., is the author of a very interesting, if highly controversial article, entitled "Philosophy of Science and History of Science."[9] In it, Clark makes the following distinction:

There are, it seems to me, at least two significantly different but basically complementary ways in which to work in the field of the history of science. The first way I call "die von unten bis oben geistesgeschichtliche Methode." I mean thereby a research policy which prescribes as its own point of departure the earliest accessible inception date of the scientific enterprise and sets as its goal the attempt to reconstruct in as comprehensive and complete detail as possible exactly how contemporary science of any given date has come to be what it is. In the vertical shaft of history this methodology thus works from the bottom to the top. It is therefore more or less compelled to organize its researches according to the structure of a logically and systematically irrelevant pattern of standardized chronological divisions or to adopt with some minor adaptations conventional but alien periodizations already established in the field of general history. One further hazard of this "von unten bis oben" methodology is that it leaves its devotees open to invasion by the precursitis virus, an affliction which differs principally from bursitis in the fact that whereas the latter causes pain to the victim and excites sympathy in the spectator, precursitis exalts and thrills its victim but pains the observer. Despite its many analytical and systematic limitations, however, this "von unten bis oben" methodology is nevertheless capable of achieving spectacular results. One outstanding example of the success of this procedure is the monumental opus of George Starton's Introduction to the History of Science. The second way in which to work in the field of the history of science I shall call "die von oben bis unten" geistesgeschichtliche Methode. I mean thereby a research policy which prescribes as its own point of departure the logically sound conviction generated by the creditable results of a rigorously analytical philosophy of science that the scientific endeavor of mankind has finally come of age in the twentieth century. . . . In the vertical shaft of history this methodology works from the top to the bottom. It therefore sets as its goal—not merely to reconstruct in as comprehensive and complete detail as possible exactly how contemporary science of any given date has come to be what it is—but rather to disclose more by analytical than cumulative procedures exactly how it happened that the history of the internal development of science is as long as it is. This "von oben bis unten" methodology is therefore free to organize its researches according to the structure of a logically and systematically relevant pattern of central ideas and to invent its own periodizations independently of the conventional framework of general history.

In his comments on the article we just cited, I. E. Drabkin states: "At any rate, as I see it, there are two diseases, not one—precursitis (to use Father Clark's term), the tendency to see continuity where none exists, and what we may call vacuitis, the failure to see continuity where it does exist."

Our goal has been to establish a few reference points on the way that leads from Aristotle to seventeenth-century mechanics, while trying not to let ourselves get infected by illnesses identified by Clark and Drabkin.

Still, we really did not worry a great deal about these illnesses, because, actually, we do not believe that the history of the sciences can be classified exhaustively according to the two complementary methods distinguished by Clark. More useful seem to us the remarks by E. J. Dijksterhuis: "the history of science constitutes not only the memory of science, but also its epistemological laboratory." Only as "memory" can the history of science be analyzed *von unten bis oben* (from the bottom up) or *von oben bis unten* (from the top down), following Clark's terminology. As soon as one looks at it as an "epistemological laboratory," one is concerned with very different issues. In the preceding sections we indicated some of the problems that appeared to us particularly pertinent. In this respect, our position differs somewhat from that of Dijksterhuis concerning the scope of epistemological analysis as part of the history of science.

In the last part of his excellent article already cited, Dijksterhuis makes reference—very clearly and pertinently—to the development of classical mechanics. He notes in particular that the Newtonian theory of gravitation is one of the best historical examples that exist showing how "every solution of a scientific problem raises new questions which are partly of a scientific character again but also partly of an epistemological nature." He illustrates his affirmation with the following examples:

The school child repeats it thoughtlessly after his teacher: the stone falls because the earth attracts it. But what is this attraction, and how does it happen? May it also be said that the stone tends towards the earth, or is impelled towards it? When the pupil has advanced a little further, he learns to say that every body perseveres in its state of rest, or of uniform motion in a straight line, unless it is compelled to change that state by forces impressed upon it. But in relation to what frame of reference does this statement apply? In relation to absolute space? If so, what is absolute space and how can we establish absolute motion? Finally the student becomes acquainted with the general principle of gravitation, and thus learns to explain the motion of the planets about the sun, that of the moon about the earth, the tides, and the motion of bodies falling on earth. But what does "explain" mean here? To what extent does that which is announced as an explanation satisfy man's desire for causality? What do we understand of the phenomenon now? Is this understanding something else than a description in mathematical terms? If so, what is this something else? If not, do we have to conclude that to understand a thing is no more than to subsume it under a general notion with which one has become familiar?

We shall see, in chapter 4, that this way of formulating the problems corresponds fairly well with the way the debate was engaged between the Newtonians and the Cartesians and between the Newtonians and the followers of Leibniz. For the moment, what we are interested in is to show that the way Dijksterhuis poses the epistemological problem encounters certain limitations, which make the analysis insufficient and not very fruitful. The historical analysis acquires another dimension, however, if we forget about the too direct questions formulated by Dijksterhuis (what do we know? what do we explain?) to go back to formulating the problem as proposed by genetic epistemology about thirty years ago. As we indicated in the introduction, the key to the interpretation of the historical evolution of a science is to ask *how* it passed from one period to the next—that is, what are the cognitive mechanisms involved at each period and which are those that facilitate the passage to a higher level? This is precisely what our present preliminary conclusions attempt to answer.

To begin with, we have already mentioned that by the twelfth century—a time when Aristotle's physics was the normative system of reference for all studies concerned with the dynamics of motion—the methodology of science had surprisingly been developed to a degree of perfection which compares favorably with the philosophy of science that is dominant today in the Western world. (Maier, Clagett, Crombie, among others, have done much reasearch in this area in recent years.) The reasons why the contents of the Aristotelian theory of motion—as well as that of his medieval followers—differs from the ideas advocated by Galileo, Descartes, Huygens, and Newton are thus not to be sought where the traditional accounts of history of mechanics have placed them. The long evolution from Aristotle to the seventeenth century, is not due to the difficulties the authors involved in this slow process might have had in convincing themselves of the advantages of experimentation or of the fruitfulness of the hypothetical deductive method. We already indicated that there was not much to be added to what had been said on that subject in the thirteenth century.

We shall relate this historical fact to a general thesis to which we shall continually return: it is not methodological considerations (or at least never these alone) that cause a science to advance at critical moments of its development, but rather the modifications the epistemic framework undergoes, which applies or directs the methodology. This framework evolves independently.

The question to be asked is, then, not how an adequate method was developed which made it possible to conceive of a theory of motion that satisfies the contemporary criteria of a scientific theory (Clark's "from the bottom up"). We should rather ask what the epistemic foundations were that underlay the applications of a given methodology, and what had been the evolution leading to the epistemological prerequisites, upon which were based, from the seventeenth century onward, the criteria for accepting a particular theory as scientifically satisfactory.

Thus, we agree with Dijksterhuis about the necessity to proceed to an epistemological analysis with the aim of determining the historical stages of a science, but we believe that it is essential to establish a distinction between aspects of methodology and the epistemic basis. Thus, the problems we raised throughout the present chapter have concerned the following fundamental issues:

a) the type of question a given theory attempts to find the answer to;
b) the type of nondemonstrated premise, explicitly or implicitly accepted;
c) the kind of relationship between experience (experimentation) and the theory;
d) the role of mathematics in the formulation of a physical theory.

Let us emphasize once again that these are not questions of methodology. These questions remain problematic even when an adequate methodology has been elaborated. (That is, once it is accepted that from experience one may obtain, by induction, "general laws"—regularities—and also employ a hypothetical deductive method using empirical observation to verify the consequences deduced from theoretical hypotheses.) The answers to such questions can, thus, not be ascertained in terms of methodological norms. They will rather come from historical research designed to discover the epistemic prerequisites that characterizes each stage in development as well as the cognitive mechanisms involved. From this point of view, we think that the development of mechanics from the Greeks to the seventeenth century is characterized by three specific kinds of "transition." On the basis of these three forms, one can adequately define the scientific revolution which took place in the seventeenth century as a result of a change of the epistemological framework.

1. The Transition of Pseudo-Necessities and Pseudo-Impossibilities to Logical and Causal Necessity

Aristotle's reasoning is logically so rigorous that once the premises are accepted, the conclusions are inevitable. These premises he presents as being necessary even if undemonstrable. Where does this necessity come from? Aristotle's reasoning corresponds to the stage which psychogenetically is characterized by the primacy of pseudo-necessities and pseudo-impossibilities (chapter 2, 11 (1)). This pseudo-necessity has various sources, among them the consideration of whatever exists as being necessarily such, and the derivation of a world view from religious conceptions.

In the former case, the two distinctive characteristics of pseudo-necessity which we found in psychogenesis emerge clearly throughout the argumentations: a confusion between what is general and what is necessary, on the one hand; and a failure to differentiate between what is *factual* and what is *normative,* on the other. There is the sense that "it has to be that way," which imposes strict limitations on what can be accepted as being possible. Such limitations remain effective for a long time, and in the evolution of scientific thought they constitute barriers impossible to transcend.

Therefore, it is incorrect to blame Aristotle for not having observed nature, it is equally incorrect to believe that such observation of nature is the primary characteristic of the seventeenth-century scientific revolution. Aristotle was an empiricist. His "errors" as a physicist are not due to any kind of incapacity for observation or to a relative refusal to carry out empirical observations. They were related to the epistemological presuppositions introduced into his "reading" of experience and, consequently, to the use he makes of observation.

Let us look more closely at the way these "prohibitions" are imposed by "pseudo-impossibility" and "pseudo-necessity." Bodies left undisturbed at a certain distance from the surface of the earth rise or fall. Earth and water descend; air and fire rise. A mixture of several elements will conform to the movements of the predominant element, since one never observes an immobile body remaining suspended in air. From this *general observation* there follows an equally general *premise:* a body cannot simultaneously make more than one simple movement. From this premise—based on observation, let us repeat—one is able to deduce the movement described by a projectile propelled horizontally: first the "forced" movement predominates, so that the projectile will follow a horizontal path until it stops; then the natural movement becomes effective, so that the body will begin to fall.

This kind of conception was to last, with few changes, till the sixteenth century. Albert of Saxony, in the fourteenth century, modified it slightly following Buridan's theory of the impetus: there is a first phase, when the impetus wins out against weight so that the body moves horizontally. When the impetus begins to become exhausted, the body's weight comes to win out against the impetus and the trajectory begins to incline downward; finally, when the impetus is completely exhausted, the body descends vertically. Even though they considered brief intervals of struggle between two forces, the "impossibility" of thinking of a composition between these forces was so tenacious that Oresme himself, in his comments on *De Caelo,* thought himself obliged to give a long justification designed to eliminate even the possibility of "compensation."

The importance of "pseudo-impossibilities" and "pseudo-necessities"—which impose limitations on the development of hypotheses and on progress in the building of scientific theory—cannot be overemphasized. The history of mechanics (from Aristotle to Newton) could be described as a history managed by the process of eliminating pseudo-necessities. The most dramatic moment in this history and the best known one was when Kepler tried to free himself of the "necessity" of planets to describe circular motions. This led to the discovery of the elliptical motion of planets. Even Galileo had been trapped by this requirement of assuming a circular motion for planets; this type of motion was considered the simplest and most perfect one.

2. From Attributes to Relations

The second aspect to consider is the introduction of quantity where before reference had been made only to quality. Now, to quantify means to compare—to establish relationships. When one replaces a discussion about properties with one about the relationship of objects to each other, one simultaneously changes the kinds of questions one envisions finding answers to in order to "explain" motion. In addition, this implies a relativization of concepts which previously had been conceived of only as absolute. Let us consider the case of a stone falling freely or in an oblique descent. While Aristotle and his followers had asked about the nature of the falling body and how its attributes are modified during the fall, Galileo puts such questions aside, concentrating exclusively on a comparison of the distances and durations of falling objects. Newton, for his part, reduces the problem to a relationship between a system consisting of earth and stone and shows, in an ingenious synthesis, that this relation is the same as that between a system consisting of sun and earth. To get to that point he also had to eliminate another limitation imposed by Aristotle's theory: the "necessity" of perfection and incorruptibility attributed to the Universe beyond the Moon, and which, therefore, obeys laws that are entirely different from those of the sublunar universe.

The absolute properties of bodies dissolve in a system of relations where, in the final analysis, one makes reference only to intervals of time and distance. This change from attributes to relations, along with a relativization of concepts resulting from it, is not exclusive to the great scientific revolution of the seventeenth century. We find it in all the important revolutions in the field of mechanics. Intervals of time and distance would become relativized by Einstein. Both enter, not as attributes but as relations, into a larger system, which also includes the "observer's" frame of reference. These higher order properties would in turn undergo the same historical process, since even speed would lose its absolute sense to become a measuring tool in quantum mechanics, at least in the version of N. Bohr.

We shall return to this mechanism operating in the conceptual evolution of physics in chapter 7. For the moment, we simply wish to emphasize the enormous intellectual effort required by each of these historical jumps; they involve nothing less than the substitution of relations of almost "tangible" properties by an abstract system.

From a historical point of view, the culmination of this process is reached with one of the most famous expressions of Galileo: "I contend that there is, in the external bodies, nothing that could stimulate our sense of taste, smell or sound other than dimensions, forms, quantities and slow and fast motion" (*Il Saggiatore,* question 48).

3. The Transition of a "Physical Explanation" in Terms of Ultimate or Concurrent Causes to the Conception of a Dynamics Establishing Only Functional Dependencies and Systems of Transformations

Even though this aspect of the problem is intimately related to the preceding point, it still does not constitute a consequence that follows directly from it. The abovementioned change from attributes to relations, certainly involves an identification of the parameters and their subsequent quantification. But here we do not speak simply of measurement, but of introducing the concept of a functional relation between variables that characterize the state of motion of an object at different points in its trajectory. This implies, first, the introduction of time as a separate variable. Galileo was the first to have done so explicitly. Thus, he succeeded in accomplishing one of the most important advances in the construction of what later became Newtonian mechanics.

In addition, the new mechanics defined force as a function of space and time (conceived as independent variables) and also introduced the concept of parameters that have constant values for each object. These innovations brought about the most profound modifications known in the history of the science of motion into the core problem of formulation. This time the credit went to Newton. His most ingenious contribution—apart from the synthesis he achieved in the mechanics of celestial and sublunar bodies—was his conception of problems of dynamics as "problems of initial conditions," as they were later called in physics: The values of the parameters of a system at a particular point in time and space (that is, precisely, the "initial conditions") determine the subsequent evolution of the system. The aim of mechanics is to calculate this evolution, without asking further questions concerning the "real causes" of motion. But the evolution itself is computed on the basis of a system of transformations which permits the substitution of the values of the variables from those assigned to them at the initial stage to those they may assume at any given moment.

This change from a search of the final cause to the construction of a system of transformations brought about a decisive advance in the history of mechanics. It became one of the most solid pillars in the scientific revolution of the seventeenth century, and has led to a profound revision in the conception of the relations between mathematics and the world of physical phenomena.

II

Psychogenesis and Pre-Newtonian Physics

The analogies we wish to investigate between the psychogenesis of knowledge and their development in the course of the history of science may be of two kinds. Our general problem is to uncover the common functional mechanisms that ensure the transition from one level of knowledge to the next. This must be done independently of the absolute value of the knowledge, since these are formative processes operating in any kind of new cognitive construction. In addition, it is sometimes possible to find certain analogies in the substantive meaning of the concepts that aid us in locating and understanding the phenomena. Once a discipline has constituted itself as a science, as physics did after Newton, there is of course no longer any substantive relationship between scientific and psychogenic concepts. The latter, in spite of the necessary role they play in a kind of embryonic way, do not yet have any of the theoretical characteristics of scientific thought. On the other hand, the prescientific conceptualizations found in the evolution from Aristotle to Newton lend themselves, precisely because of their prescientific character, to instructive comparisons with the cognitive constructions proper to psychogenesis. We shall now make such comparisons of contents.

Following this, we shall attempt to analyze certain common mechanisms relating what will have been shown for this prescientific period of physics to the processes inherent in psychogenesis. Specifically, the search for common mechanisms will concern the mechanism of pseudo-necessities, as well as the sequence of predicates, relations, and transformations. In addition, we will attempt to give a psychogenetic interpretation of the discrepant relations analyzed above between the surprisingly highly developed precision in methodology of the thirteenth century and the relative paucity of scientific results found in the same period.

I. THE PSYCHOGENESIS OF THE NOTION OF IMPETUS

1. Preliminary Remarks

To present schematically the history of the notion of impetus, ignoring the accelerations and regressions that may have occurred, we can distinguish four main periods: (1) the Aristotelian theory of the two kinds of "movents" prevails, which accepts, in addition to external causation of movement, a force endogenous to the mobile; (2) the internal "movent" is no longer invoked in the sense of the preceding period and movement is seen as a global driving mechanism, without the distinction between what later becomes force and impetus; (3) the impetus (or élan)* is now seen as resulting from force and produces motion, thus occupying an intermediate position between the necessity and the cause; (4) finally, the impetus becomes the result of motion caused by force and sooner or later comes to be equated with acceleration.

Now, the equivalent of these four periods can be found in the course of psychological development. What is more, the child spontaneously formulates an original idea reminiscent of the impetus; she even borrows the term "impetus" from adult vocabulary, but her meaning is far from an adults. Such convergence may appear as fantastic and rather implausible. Therefore it seems important to show that it is in fact quite natural. There is, of course, no question of appealing here to any kind of ontophylogenetic parallelism in Haeckel's sense, for obvious reasons: (a) there exists no known instance of a genetic transmission of ideas; (b) there is

*T.N. The French "élan" has no real equivalence in English. The best translation in the present context would seem to be "impetus."

no line of inheritance between Aristotle or Buridan, for example, and the young Genevans or Polish children who served as subjects in our observations; (c) most importantly, the child's mind is logically prior to the adults of history, who of course once were children themselves.

At any rate, any convergence that may exist between "theories" of our scientific past and certain constructions, easy to follow in psychological development, remains a correspondence between cognitive products at two very different levels in a hierarchy of thought structures. The theoretician formulates problems, and in solving them makes use of concepts, inferences, and operations that are more or less clearly "thematized"—that is, have become cognitive objects. This thinking is evidently reflexive, since the search is theoretical in nature (this does not prevent the theoretician from instrumentally applying operational processes that are not yet thematized but will be so by the next generation). Furthermore, this reflection is always dependent on previous work.

Children, on the other hand, generally do not normally pose the questions which we ask them to respond to. If they happen to have done so, it is only momentarily and not with a reflective view (theories begin to be developed only at adolescence). Their answers to our questions are thus taken from the implicit interpretations, in an effort at becoming conscious, which they have developed in the course of their actions. But these interpretations exist, since in order to produce, predict, or explain movements it is obviously necessary to organize the various aspects and to give them meaning. As practical and utilitarian as these meanings are, some of them are conceptualized, as can be seen from the spontaneous use made by children, beginning at age seven or eight, of the notion of "impetus," which before is invoked rarely if at all.

In our comparison of the history of the concept of impetus with its psychogenesis, we shall establish correspondences between two developments situated at two very different levels. Their interrelations, however, become intelligible if one considers a basic law of cognitive construction: this is because, as we have noted in the introduction, the different constructions do not simply follow each other in linear fashion; they give rise to reconstructions of what precedes and are integrated within the following stage. Therefore, the ideas articulated at a higher level of thought, even those of a theoretician, must necessarily rest on substructures of actions which sustain them and which simultaneously become enriched in various ways. Just as categorical axiomatization (an extremely contemporary form of reflection) works on intuitive concepts, which it reconstructs by formalising them, any kind of reflection proceeds by reorganizing content from lower levels—in some cases going back all the way to the action schemata. Now, everything related to motion and force, including the particular form of "impetus" children talk about (in relation to their own speed or that of vehicles or mobiles they happen to use) gives rise to the formation of such schemata, which constitute the substructure of thinking relative to these notions. Therefore a territory exists which common to all subjects whatever their intellectual level, so that there is nothing implausible about the correspondence between the development of these schemata and that of the reflective ideas themselves, even though the latter go far beyond the level of the latter.

This is of course true, since, as we have seen, the level-by-level reconstruction always leads to new constructions, through extension of contents and enrichment of structures. These innovations are, in turn, the product of mechanisms that are functionally repetitive, so that if our hypotheses are true, what happens at the more advanced levels in the scientific history of the notion of impetus is evidently the result not of simple transposition into the language of reflective theories, of what had been elaborated at the more elementary levels, but of a formative mechanism, which relates the four phases in the reflective interpretation of motion, analogous to the way the four levels of practical, later symbolic interpretation of motion and its causes are related to the earlier stages in children's conceptual development.

What is surprising, however, and might pose a constant stumbling block, if we did not have any information about its nature, is the difficulty in coming to conscious awareness. For example, why is it that the central role of acceleration was discovered only with great difficulty, although we are all in a position to experience it through the movements of our own bodies? Similarly why is it that all effort constitutes a manifestation of this acceleration rather than, as was thought by Maine de Biran, of a kind of absolute "force" that depends only on the subject's will or the capacities of the organism? (This is the evident source of the "internal movement" of Aristotle.) Or, in another domain, why is it that Aristotle's logic, born of a reflection on everyday reasoning, has not achieved consciousness of the structure of relations, which are used just as currently as the syllogistic structures? Many other examples could be given, all showing that there may be considerable delays between the reflective construction of a concept and its sometimes even systematic use at the lower levels of action schemata or of already symbolic, not yet thematized substructures (that are not even susceptible to thematization). Thus, it is entirely natural and understandable to find correspondences between the constructions in psychogenesis, which subsist in every normal subject at levels below the most advanced forms of thinking and historical stages.

2. Stage 1

As to these correspondences with history, it seems clear that concerning Aristotle's theory of the two kinds of motor, the principal issues are (a) those concerning the necessity of assuming an internal motor, even in cases of mobiles other than animate beings and (b) the necessity of a continuous contact with the external movement in cases where the conservation of imparted motion would seem to be sufficient.

Concerning the first point, the comparison with psychogenesis seems illuminating, as it shows the relatively late development of the differentiation between living beings and inorganic substances. From this point of view the generality of the internal motor is self-evident and closely related to that of teleological causation, which today is associated exclusively with the biological domain, where it has the causally elaborated status of teleonomic regulation. Aristotle, on the other hand, associated all motion with a biomorphic goal, as is also true for the elementary stages of thinking. In the child, the hypothesis of an internal motor goes quite far. Very young children, for example, may believe that the wind is produced by trees (which sway by themselves), by waves that rise, or by clouds, which spontaneously move ahead. This naturally favors the formation of the antiperistatic schema, to which we shall return.

As for the necessity of permanent contact between the external movement and the mobile, the two reasons that are decisive for Aristotle also play a larval kind of role in psychogenesis. The principal reason is, as we have seen, the failure to accept any kind of inertial motion, which is all the more curious in the Stagirite. There, by very accurate reasoning, Aristotle demonstrates its necessity in the case of the void, but from this deduces only the absurdity of both the void and of inertia itself. The second reason is the theory of the proper or natural place, according to which a projectile should fall to the ground in a vertical path, once it is separated from its propelling device (which both Aristotle and young children believed is the case).[1] Now, we find that children formulate these reasons more or less explicitly, which frequently and systematically leads to antiperistatic representations. As for the lack of belief in inertia, it seems self-evident that a rectilinear motion should be thought as coming to a stop because of some resistance, by a simple loss of "impetus," or by fatigue. But when the movement is continuous and without an apparent initial propellent, as in the case of clouds, subjects frequently say that the clouds are propelled by the wind, which they produced in the first place by moving just a little bit: this brings about a "surrounding reaction," which may become more or less powerful. In the case of centrifugal force, another form of inertia, young children will say that a solid or liquid does not fall when it is at the peak of the circular path described by its recipient, which "makes air" by rotating. This air both pushes and retains the heavy body near the bottom of the container.

Many other examples of this retroactive shock imparted by the air could be cited in situations where it would be least expected, often becoming the explanation useful for all cases in which an unexpected effect is produced. Therefore, it makes good sense to ask children a question about projectiles: why is it that a ball does not fall down as soon as it is thrown? Young children do not see any problem there, because one "throws hard" and because it is made for such movements (internal motor), but older children will invoke Aristotle's idea of the air being displaced by the propellent and the air current produced by the mobile which comes up behind it.

As for proper place, clearly, the pebble belongs on the ground, while smoke belongs in the air. No connection is made between their relative weights.

It is interesting to cite some of the spontaneous remarks made by the subjects: one child of seven or eight said, without being questioned about the reasons for the descent of a marble, "If we put it here (departure), it certainly needs a place it always goes to; it must have its natural place." It is hardly possible to be more peripatetic about the fall of bodies.

3. Stage 2

The second stage is characterized by the disappearance of the internal motor; at this point, a certain number of powers are bestowed on the external movent. However, they have not yet differentiated but include a number of aspects, which later come to be distinguished under various names corresponding to the force of the agent, the impulse received, velocities, distances, and mass (in collisions and resistances). Such a large polyvalent notion is reminiscent of the variable "action" in physics. Now, it is well known that Newton did not formulate his first law according to the equation $f = ma$ due to Euler, but under a form corresponding rather to $f = d(mv)/dt$, which refers to action. But since we are dealing here with metric infinitesimally graded, differentiated variables while the primitive "action" we are talking about here can be broken down only into qualitative concepts, decomposed since it remains undifferentiated, we shall call it "global action" and write it as *mve*.

Given this, the experimental situation in which one can best discriminate the characteristics of this second stage, in which the impetus is not differentiated from the external force (as opposed to the third stage at which that force, the impetus that results from it, and the movement produced by it are dissociated) is that where the transmission of motion is seen as mediated by immobile elements.[2] A marble hits a block and initiates movement in another marble on the other side of and contiguous with the block, or it may hit the first of a row of marbles only the last of which will be set in motion. In stage 1, subjects still appeal to an internal motor: the marble hits the block, which remains immobile; but the marble on the other side starts by itself by a kind of contagion and by its own force. In the case of the row of marbles, the active marble passes behind the others without being seen, and pushes the first marble, whose place it takes, once the first has left its position, or it goes back to its own place, etc. In stage 3, subjects say that the active marble, because of its force, gives an impetus which passes "across" the intermediate elements and provokes the passive marble into motion. But in stage 2, the internal motor is eliminated, but the intermediate impetus is not yet differentiated and the active marble is seen as the source of some "global action," which works as follows: (a) First, there is a push $p = mv$, determined by the velocity v and weight m. (b) This movement v is transferred to the object hit (the block or the first immobile marble), but (c) the latter also possesses a weight m, which offers resistance; hence the active marbles motion is stopped. Let us note that in this case, resistance is seen as subtracted from the action of the marble in motion, as for Philopon, and there is no question of proportionality. (d) On the other hand, when the object hit receives the push, it must advance, and this push $p = mv$ cannot be dissociated from a distance covered e within the global action mve. Now, this is contrary to what can be observed, since the block and the intermediate marbles remain immobile: but the complex mve is so resistant in its undifferentiation that subjects believe that they have seen the intermediate elements move when in fact they have stayed in place; and when they are asked to push with their finger on the block to keep it immobile, they affirm their impression, having felt a translatory motion. (e) This push presumably causes displacement and is seen as being repeated from one marble to the next until the last marble, which is the only one that moves because it does not encounter any resistance.

It can be seen that the global action mve leads to the perceived impossibility of mediated transmission. Instead, subjects perceive a series of immediate transmissions, with translatory motion assumed from one object to the next.

4. Stage 3

In stage 3, the interpretation becomes quite different. An "impetus" is now invoked as an intermediary between the push of the active marble and the movement of the last passive one. This "impetus" is not simply a verbal notion, since it represents a new fundamental property—that of traversing the block or the passive marbles up to the last one. Undoubtedly, these subjects still believe that the intermediate marbles need to move by a small amount to transmit the impetus received. Thus, the transmission, now intermediate, is as yet only semi-internal. But the essential fact is that from the active marble, and from one marble to the next, the push is seen as causally provoking the movement of the mobile only under the condition *sine qua non* of transmitting to it an "impetus." The latter is thus the necessary causal link between the push and the acquired motion. In this sense, one can compare these explanations to those of Buridan, for whom the impetus is the indispensable intermediary between the force of the external motor and the translatory movements resulting from the push. The analogy goes even further to the idea of possible cumulative summation of impetuses. In fact, Buridan interpreted the cases of observed accelerations by assuming that if the external force continues to be effective, the successive impetuses are added one to the other. Now, in the case of a row of marbles, where there is no acceleration, subjects often state that each "impetus" transmitted from one marble to the next is added to the preceding one so that the last marble receives a total impetus which is greater than the one the first marble received.

Here are a few examples.[3] Aug (8;0) says that the active marble gives "a good push," which is transmitted from one marble to the next. At first, he thinks of a cumulative effect, "it always pushes faster on the others"; but later he believes that the final impetus is the same as the first impetus ("the same amount of impetus"); somewhat later, he hesitates and extends the row to increase the resistance: with more marbles, "it helps to give more force" or alternatively, "it will go slower." Dan (8;4) similarly says that the active marble has "sent its impetus" to the first, which passed the impetus directly to the second," etc. He concludes that the impetus "makes a current." Per (7;8) speaks of "bumpings," which always give more force to push the last marble (cumulative effect), and this force "goes into the middle of the marble, then goes through that one, then that one . . . (etc). Web (8;4) predicts that "the impetus will go through the balls" (but then regresses to an explanation in terms of antiperistasis). Ric (8;4) believes that "the impetus gets weaker" with further resistance, but he later considers it as being constant.

It can be seen how general is this notion of a necessary intermediary between the external movent and the motion of the passive mobiles. Let us further note that certain subjects believe, like Buridan, that there is addition of individual impetuses in the case of acceleration or a weakening caused by the cumulative resistance. Others think, like Oresme, that the impetus can diminish through a kind of expense of energy (comparable to what in contemporary language might be called the cost of labor). Thus, Mon (12;8) says, comparing the first and the last of the passive marbles: they receive the "same impetus, perhaps a little less (at the end) because the others keep (retain) some of the impetus from the shock."

5. Stage 4

While the impetus was seen as the cause of motion and velocity during stage 3, the relation is now inversed. The impetus is now regarded as resulting from velocity or, more precisely, as being one aspect of it. This then tends toward the notion of acceleration. Thus, as soon as subjects witness the mediated transmission, there are some 11- to 12-year-olds who say "it's because of the speed, and because there is more and more speed, because the marbles (in the row) transmit it to each other, there is more impetus" (Ogi, 11;11). "The force is transmitted from one marble to the other.

"Where does it come from?"

"From the difference in height (the slope which the active marble follows) so it picks up momentum, speed."

"Can you tell them apart?"

"The force gets stronger with the speed, the impetus, that's the point of departure (he indicates the descent) of the speed, the speed transmits the force and the automatic impetus."

"How?"

"Through the axis of the marble" (Bia, 15 years).

In another study[4] in which a marble hits other marbles held by a horizontal rod, subjects have to indicate with a marker the respective variations of speed and impetus. The subjects of this stage, in spite of their variations in vocabulary, are unanimous in marking acceleration beautifully and in considering it the decisive factor: for Tri (12;2) the force depends on the weight ("the heavier it is the more force there is"), but it also increases with "the speed it has: the momentum. The greater the impetus, the more force there is."

Gil (13;0) imagines a very regular acceleration and talks about impetus and speed: "Where does the impetus come from?"

"The more it goes down the faster it goes."

"Is it the speed that gives the impetus?"

"Yes. Or the impetus that gives the speed. If there is no speed, there is no impetus, and if there is no impetus, there is no speed."

Impetus is necessary "because one cannot start all at once." Thus, the impetus appears as confounded with acceleration, while force also requires what he calls weight.

The decisive fact at this stage is, then, the discovery or the explanation of acceleration. If in talking about it subjects still distinguish impetus and motion qua speed, the former no longer represents anything but the increase in the latter. It no longer constitutes a separate causal factor generating the movement. The essential progress accomplished here is in an effort at quantification. While they do not yet make use of a metric proper—that is, measurements of time and distance—but restrict themselves to evaluations in terms of more and less, subjects do mark, without any ambiguity, the increases of the spatial intervals the marble runs through, holding time constant. The relation between movement and impetus is thus characterized by quantified variations in speed, even though still badly expressed verbally. As for force, it is seen as a function of this acceleration but also of mass, called weight. Still missing is a reflective system of stabilized concepts integrated within a proper theory. However, all the elements that would be necessary for the elaboration of such a theory, as far as its content is concerned, are already present: what is missing is, then, only an adequate organization and systematization.

6. Conclusions

We shall now try to describe the way the four phases just distinguished are constructed in order to compare this to the other construction mode, clearly of a much more advanced level but still definitely analogous—that of the historical formation of reflective theories, as described above.

This mode of construction can be summarized in terms of two essential processes: one is the gradual differentiation of variables during the first three stages, which results from attempts to set up correspondences and, even more importantly, from corrections of those that turn out to be inadequate; second is the quantifying integration of elements in stage 4, which once differentiated require the formation of a new structure.

Stage 1 can be characterized, from this point of view, in terms of maximal undifferentiation due to illegitimate correspondences between living and inanimate beings. The representation of motion is certainly due to a reflection of the action schemata onto the level of concepts. These schemata consist in displacements of objects by manipulation as well as displacements of one's own body. This reflection in itself involves setting up correspondences, and from this are derived a whole series of further such correspondences, which relate actions of objects to those of the subject: this is facilitated since the first instrumental or pre-instrumental behaviors at the sensorimotor level (pulling on a cover to reach a distant object, using a stick, etc.) employ intermediaries as extensions of the body. As a consequence, for these subjects, every movement has a goal, that is, subjects may assume an internal action in the mobiles or they may understand force in the Aristotelian manner as the triggering of this action rather than only in terms of an external push, etc. Hence, we find generalized this kind of biomorphism which manifests itself in the conception of a bipolarity of external and internal movents.

The first important form of differentiation appears at stage 2, when the illegitimate correspondences are eliminated, and consequently also the idea of an internal movent. The mechanism accounting for this correction is to be found in the setting up of more accurate correspondences between the movements, their antecedents, and their consequences. For example, when a marble hits a block and causes the marble on the other side to start moving, subjects no longer say, if this phenomenon repeats itself several times, that a different intermediary has taken the place of the block, or that the other marble ran off by itself like an animal that runs away from another. Rather, they interpret this correspondence as a "function" in a stricter sense—one involving a dependency. From this comes the idea that the block must have been loosened, etc., and, in general, that a push is enough without the triggering of an internal motor.

But now the external motor, the only heir, on the physical, inorganic level, of the corrections and differentiation responsible for the elimination of the internal motor, is now charged with all the power. It remains in turn, the center of this undifferentiated complex which we have termed "global action" *mve.* This remaining undifferentiation manifests itself in particular in the remarkable fact that subjects infer from the movement of the last of the passive marbles that the process must be the same as in the case of immediate transmissions, when *A* pushing against *B* causes *B* to start moving immediately. This is how subjects get the idea that each of the passive marbles makes the adjacent one advance. This interpretation is so resistant that if the experimenter makes a mark at the position of each marble, the children will answer that the marble did in fact move, but then went back to its place, after having finished pushing the other one!

Next are the differentiations characteristic of stage 3. These are maximal and precede the integrations required at Phase IV. Now, subjects distinguish among velocity, impetus, and force. But, even though they perceive a fairly general bidirectional causal relation, according to which the impetus is due to force and generates motion by means of velocity, some unsettling ambiguities make their appearance as soon as it becomes necessary to specify the variations in terms of quantity (naturally, without asking questions of a metric order). Thus, responses vary among subjects (and even within subjects), as to the greatest impetus, with respect to a slope or a row, whether it occurs at the beginning, middle, or end. A similar variability is observed with respect to the question of maximal speed. Subjects do not establish a constant relation within the sequence of events nor between the variables, speed and impetus. As for force, it depends both on the slope, hence its relation to the impetus, and on weight. But mass also is seen as variable, as a function of a relation, which one might symbolize as $m = p/v$—that is, weight is thought to increase when there is a push and decrease with increases in velocity: we can conclude that there is definite progress in the differentiation of variables or factors, although due to a lack of coherent quantification, they are not integrated within a stable system.

Stage 4 marks a point of relative equilibrium due to a quantitative integration. This is all the more remarkable when one considers that there are still no measurements so that the synthesis does not go beyond the level of increasing and decreasing functions. Integration manifests itself in two ways. First, there is the empirical discovery of acceleration on the slopes, with a structure of spatial intervals, which increase within a constant time interval or remain identical with shorter durations. This permits the union of the impetus with velocity, the former becoming merged with acceleration. There are even some 11 and 12-year-olds who talk about increases in speed, defining the impetus as a uniform increase in velocity. Secondly, this structuring is accompanied by the development of three kinds of conservation: weight is no longer considered to vary as a function of the force of the initial push or of speed; subjects no longer see cumulative effects or losses of energy in the transmission of movement across the horizontal row of marbles (this movement is now thought of as entirely internal and not involving any translatory movement); this achievement is a first sign of the conservation of quantity of movement and kinetic energy. Finally, subjects can see the equivalence in height between the points of arrival and departure in a situation where marbles descend a slope and go up again on a symmetrical or any other type of gradient.

It can be seen that the differentiations of stages 1 to 3, which were due to corrections and improvements in the correspondences, lead to the formation of new structures necessary for the coherent integration of the variables thus differentiated. But these structures are not the product only of correspondences: they also require the use of quantifying operations, which are both transforming and conserving and which are formed at about 11 to 12 years, as we know. Let us note in this connection that the discovery of the "impetus that goes through" the marbles made at stage 3 already involves operational transitivity, here causally "attributed" to the mobiles themselves.

Given this account of the psychogenesis of the notion of impetus, there does not seem to be anything implausible about finding corresponding historical periods during which the notion of impetus has arisen and led to a concept that became the origin of the notion of acceleration as an essential component of force. There undeniably exists a considerable difference in the structure of developmental levels between a series of reflective theories and elementary levels in the growth of consciousness and conceptualization, where simple action schemata become internalized as operations. But in both types of development, thinking can advance only by using analogies and identifying differences while finding increasingly refined correspondences followed by integrations within quantitative structures that arise through operational transformations. The early development of these functional mechanisms, which can also be found later at all levels, should not be seen as indicating that scholars, from Aristotle to the Pre-Newtonians, had drawn their ideas from their childlike unconscious. If that was true, there should have been an immediate extension rather than similarity in the constructive processes. Rather, the similarities encountered show something far more interesting—i.e., that intelligence functions in the same manner at all levels. At each level, it must be reconstructed so that it can transcend previous levels. Epistemic progress does not consist in simple additions, but rather in new forms of organizing knowledge. These lead to the creation of new knowledge. There is, however, something mysterious about the sequence of stages in the development of the notion of impetus, which occurs much more rapidly in child development than in historical development. The reason is undoubtedly related to the influence of the adult social environment manifesting itself in countless stimulations and ever new problem situations. But this does not mean that the children's responses had been dictated by simple learning. The fact that the surrounding intellectual climate should stimulate children in the direction of quantification only raises new questions; it is up to the subjects to build their own cognitive tools in each new problem situation.

II. THREE COMMON MECHANISMS

1. Pseudo-Necessity[5]

This first form of cognitive response—even though it naturally leads to certain conceptualizations—definitely concerns a "mechanism" and does not characterize epistemic "contents." It is a process which is certainly common to psychogenesis, where it plays an important part, and to the history of prescientific thinking, and of sciences of a more advanced level as well.[6] It is peculiar in that it functions as an obstruction, and the corresponding constructive mechanism consists in a "liberation from pseudo-necessities" (this may be relatively rapid or fairly slow). More specifically, pseudo-necessity constitutes an elementary phase of nondifferentiation in a general process, which leads to the differentiation and correlative coordination between the possible, the real, and the necessary. At an advanced level of thought, when cognitive activity is directed at transformations and no longer only at predicates or simple relationships, an actual transformation will have appeared as one among others possible within a system in which the compositions appear as logically necessary. But this final outcome is the product of a long process of differentiations and integrations, and the initial state is naturally undifferentiated in the sense that an observed form or movement will appear to the subject as the only one possible and thus as necessary: of this order is, thus, the "pseudo-necessity" resulting from an initial lack of differentiation between the general and the necessary, between the factual and the normative (if an object x is what it is, it is necessarily so), or again between the perceptually "good forms" (straight lines and circles, etc.) and forms that are the only ones to be rationally intelligible, etc.

Pseudo-necessities can have two kinds of effect on the reading of empirical data and on the formation of epistemic theories: they limit the kinds of possible observables and they may lead to faulty generalizations and cause these to have more weight than the correct ones. This explains the paradoxical nature of Aristotle's physics, of which nothing has survived today in spite of its perfect logical coherence and in spite of his having started out with the clearly empirical intention to base his study on facts: Now, even if these initial facts may be exact on the points observed, they are somewhat particular and limited—that is, they are deformed by pseudo-necessity so that the generalizations derived from them are both indisputable, from the point of view of their logical form, and quasi-systematically false with respect to their content, where the limitations appear in this case as deformations.

To return to psychogenesis, this process of pseudo-necessity can be found in every domain, and in all possible ways. With respect to geometrical forms, a square rotated 45 degrees is no longer perceived as a square but as a diamond; indeed its sides no longer appear to be equal. Similarly, a scalene triangle is no longer seen as a "real" triangle. As for kinematics, horizontal and vertical movements are primary and young children work with a model like the Aristotelian one, in which a projectile follows a horizontal path all the way to its goal and then descends vertically to land on it. We may add examples of other fictitious accounts due to the same pseudo-necessity: when playing a game of tiddlywink—(pressing one's finger on the edge of a token lying on a carpet, thus causing it to leap into a high-sided box), children can obviously see the curved path taken by the token, but they interpret it as follows: the token slides horizontally across the table, and when it arrives near the box, it jumps up vertically to pass over the wall of the box. In the domain of causality, we find that all young children declare that the water in the river runs down "because it has to go into the lake." A small boy told us that the moon does not shine in the daytime but only at night, because "it is not in charge"—a telling example of nondifferentiation between the factual and the normative, reminiscent of the cosmogonies of Antiquity.

In short, "pseudo-necessity" is a common phenomenon during

the early levels in the psychogenesis of knowledge. The difficulty lies in imagining possibilities other than the ones that actually occur in a particular instance. As such it clearly represents an initial nondifferentiation among reality, possibility, and necessity.

The following phases are then characterized by the opening up of new possibilities and the constructions of other necessities. Now, as future possibilities are not predetermined and each time require a new construction, two consequences naturally follow. The first is that pseudo-necessities are all the more frequent as knowledge acquisition occurs only in the initial phase, in which it is very difficult to evolve new possibilities given the limited number of elements known that could be combined: this explains the analogies between the pseudo-necessary inferences found in psychogenesis and those which persist in so obvious a fashion in Aristotle's physics, which was a first attempt at creating a complete system of the knowledge of mechanics of his time. The second consequence is that, at all levels of scientific thought, the discovery of a new possibility may in turn be prevented for a long time by these pseudo-necessities. Let us cite a well-known example: it is not without reason that the curves without tangents were a late discovery (by Bolzano and Weierstrass): geometric "intuition" erects a barrier of pseudo-impossibility against such a notion, which is particularly powerful and resistant.

2. Predicates, Relations and Transformations

The second general mechanism demonstrated is the change from a centration on predicates to a consideration of relationships and later transformations. This is certainly the most important of the mechanisms common to the history and the psychogenesis of knowledge. It may have other forms and different contents than the particular case of pre-Newtonian physics; in general, it corresponds to the three important stages, which we have called "intra," "inter," and "trans" in our introduction.

With respect to psychogenesis, this is a very general, and necessarily sequential process, since in order to attain the level of transformations one has to go through those where relations and covariation are treated. These in turn, even in their simplest form, require as a prerequisite an analysis of properties or predicates. The necessity to go from such analyses to one involving relations and transformations derives from a fundamental property of physical knowledge—that it cannot be reduced to a collection of observables, but can advance only as a result of subjects' activities, which make them vary and thus subordinate the observables to endogenous systems of coordinated actions.

Since here we no longer compare the contents of development, but wish to expose the mechanisms as such, the most general illustration of the developmental law involved is the way new possibilities are opened up. This illustration has a further advantage: it extends directly what we have just described concerning pseudo-necessities and pseudo-impossibilities.

In order to study the formation of possibilities at different levels of development,[7] we have presented children with various kinds of tasks: (1) asked them to combine, say, three dice on a surface in "all possible ways" or to imagine "all possible paths" between two points; (2) to imagine all possible shapes an object might have, of which only a small portion is visible; (3) to section a square surface or other shape in all possible ways; (4) to articulate in all possible ways a chain of interconnected rods; (5) to find as many ways as possible to reach a particular goal (raise the level of water by immersing various objects, etc.) (6) to construct as many triangles or other geometrical forms as possible out of a series of sticks; etc.

Now, all these different situations have provoked remarkably comparable responses with respect to their order of appearance. These, given the very open character of the questions posed, are susceptible of shedding light on the underlying reasons for the process predicate → relations → transformations since, particularly in the sphere of prescientific thought, with which this chapter is concerned, this process is strictly subordinated to the formation of possibilities, but also inhibited by the resistance offered by pseudo-necessities.

Generally speaking, we have observed three main stages as follows: (a) a stepwise process of possibility formation by analogical succession, based on the properties of preceding exemplars; (b) anticipation of co-possibilities and their interrelations; and (c) unlimited series of possibilities, subject to the laws of recursive generation.

Analogical possibilities are characterized by a remarkable scarcity of variations from one actualization to the next. Each new variation is only minimally different from the preceding one, to which it bears a great deal of resemblance. New possibilities are generated one at a time, even though the series becomes quite long under the impact of repeated questioning: "Can you do it another way?" This, then, is a situation which is completely dominated by pseudo-necessities, where the principal instruments are qualitative predicates, with the relational aspects that are visible to the observer remaining completely implicit for the subject.

When subjects reach the age of seven or eight, one observes a notable change. They now anticipate several possibilities at once. These become co-possibilities, as they entertain explicit relations to one another: a path between two points can be straight or curved, sinusoidal with arcs varying in number and shape or sawtooth-shaped with various angles, etc. The first co-possibilities are still limited to those realizations the subject actually produces. After the age of nine or ten, these are considered only as examples among many other conceivable ones, but which subjects do not attempt to realise in their entirety.

Finally, at 11 or 12 years, the notion of possibility comes to include "anything whatsoever" in intension and unlimited numbers in extension (an "infinity," as one subject says). This new concept also includes a further innovation—the variations generating new possibilities may be so minute as to be invisible, as the subjects frequently point out. That is, they follow a recursive law—for example, the one that generates all the points on a line, etc. Some subjects even accept that the "infinite" number of possible changes of the position of three elements on a carton are the same on a large carton as on a small one: thus, it is clear that at the third stage, subjects transcend by far the level of observables in the direction of an operational system of transformations. It is true that in this particular example, the transformation consists in the simple law $n + 1$, but when applied to the formation of possibilities, it takes on a more subtle meaning than a simple series of numbers. As for more complex transformations relative to problems that are more strictly in the domain of physics, the next chapter will provide examples from psychogenesis.

3. Methodology and Epistemic Frame of Reference

The conclusions of the preceding chapters draw the epistemological lessons from the strange fact that was discovered by recent work: as early as the beginning of the thirteenth century, when Aristotelian physics still dominated the scene, a remarkably precise methodology due to R. Grosseteste, Albert le Grand, Roger Bacon and others had already provided very precise analyses of the conditions for induction and experimentation as well as of the hypothetical-deductive method. It was, thus, not the advances in methodology that led to the development of seventeenth-century physics, substituting well-established experimental facts for Aristotle's incomplete and inaccurate data, but the discovery of new problems and the transformation of the epistemological framework. Two questions remain to be answered, which require us to make a comparison with psychogenesis. First, why is it that these transformations in conceptualization did not follow directly, during the Middle Ages, the constitution of this methodology? Secondly, and more importantly, how was it possible to construct such a methodology long before any detailed application was made of it?[8] That would be the equivalent of compiling an *Ars Poetica* before any poetry was written.

Now, the comparison which we are about to propose may appear utterly fantastic at first sight. But it turns out that, as we studied, with B. Inhelder,[9] the induction of simple laws of physics in 11-year-old pre-adolescents, who had received absolutely no academic instruction in the subject, we have been able to observe the formation of a methodology derived only from their logical reasoning rather than experimental expertise or pre-existing theoretical knowledge. Therefore, we believe that the facts we shall now present demonstrate the possibility that a methodology can be elaborated as a kind of applied logic. This is true even where subjects deal with problems which they have not themselves invented and independently of the epistemological frame of reference, of which the subjects had absolutely no knowledge.

Let us take an example. We asked subjects to identify the factors responsible for the flexibility of horizontal rods, differing in material, length, thickness, and form of section, in a situation where various weights were attached to the ends of the rods so that these were tilted to different degrees (measured by comparison with a standard water level). At the pre-operational level, at five to six years of age, subjects describe only what they see, without following any order or method; the only explanations produced are pseudo-necessities, sometimes quite explicit (for example, a 5;5 year-old when asked "why does it touch the water?" answers "Because it has to"). From seven to ten, subjects identify the functional relations correctly when the experimenter presents pairwise comparisons that vary only one factor at a time. However, when asked to find by themselves the variables which affect the tilt of the rods, these subjects confuse all the factors involved. A nine-year-old subject, trying to prove the effect of thinness, compares a rod that is long and thin with one that is short and thick. Then we ask him to compare two rods of the same length but differing in thickness and ask which of these two comparisons is more conclusive: he answers, without hesitation, that it is his first comparison, because (in adding the two variables) the rods "are more different." At about 10 years, subjects discover empirically that the effects of two individual factors can compensate for one another, but they still do not try to find other compensations by combining other kinds of variation.

At the level of hypothetical-deductive thought (from 11 to 12 years on) a series of innovations alters the entire situation. In the first place, subjects, before going on to conclusive experimentation, insist on drawing up an inventory of factors (through both explorations and inferencing), conceived as hypothetical, and which they feel might play a causal role. After this, they go on to test the hypotheses by varying the stipulated factors. What is remarkable about this is that these subjects apply the rule "other things being equal"—that is, they modify one variable at a time and make sure that all factors except the one being tested are equivalent for the two rods compared. Because of this, and also because they are able to state explicitly that varying more than one factor at a time makes it impossible to prove anything, the subjects naturally recognize the possibility that effects may be commutative: "Find the rod that bends the most? I would choose one that is round (in cross section), thin, long, and made of soft metal (flexible)." Finally, they come to achieve compensations and to explain them; these are of a variety of types, combining the different factors in a variety of ways. In other experiments, where only one factor is determinant, subjects arrive just as easily at excluding irrelevant variables as they do in making the correct choice: for example, they come to recognize that the frequency of oscillation of a pendulum depends exclusively on the length of the rod, and that weight, amplitude, and the force with which the pendulum is set in motion are to be excluded.

Given these facts, it is difficult to deny that these subjects apply a methodology. The problem is to explain its formation, since, as we said before, these were subjects who had not received any kind of formal teaching on the subject nor did they have any personal experience with experimental methods. The only answer possible appears to be that, on the basis of the propositional operations constructed, such as conjunction, implication, and exclusive or nonexclusive disjunction, which enable subjects to reason about simple hypotheses and to evaluate these by deriving from them logically necessary consequences, the subjects then apply this logic to the problems we present them with. In addition, these deductive operations implicate a "set of parts" and a combinatorial logic, which permit a series of analyses foreign to the simple "grouping" of concrete operations. In other words, their methodology is nothing more than logic applied to the experimental data.

But if this is so, why is it that the subjects do not construct a theory of physics? And why, above all, did the great logicians and methodologists of the thirteenth century not elaborate one more scientific than that of Aristotle? As for our pre-adolescents, the answer is simple: (1) The problems were formulated by us and not by them, and it is the invention of scientific problems that leads to the application of methodologies. (2) The questions we asked concerned only facts and laws, thus remaining at the level of induction (the inductions consisted precisely in deductions applied to the facts), whereas the active search for "reasons," that is, epistemic significance is the principal motor for the constitution of a science. (3) Whenever the subjects concentrate spontaneously on questions of causality, they succeed very well in building valid models, as will be seen in chapter 7, but only at the level of actions or their conceptualizations, without looking for a general system.

With respect to the medieval logicians and before these, Aristotle (although the inventor of logic failed to apply it to experimentation) the problem is much more general. This is all the more true as we consider methodology as an "applied logic." In this way, we must interpret causal explanations as being based on the operations of the subject applied to the objects and their material interactions. Thus, it should be easy to go from application to causal attribution. Actually, this is not at all true. The reasons for these difficulties will have to be sought once again in the relation between reality, possibility, and necessity.

Reality consists at first of observables that are directly attained by perception; thus, people generally believe that they know something. When a fact repeats itself with some generality, it is therefore conceived as being the only necessary and the only possibility in its domain. In order to feel the need to verify that it is really what it appears to be, a prerequisite condition is to go beyond reality and to imagine other possibilities, and consequently to invent problems about issues where none present themselves. This is where pseudonecessities arise, and these can be so constraining that one might interpret Aristotle's physics as their renowned victim. The only possibilities envisioned by him did not refer to systems of operational compositions but to those predetermined processes which he called transitions from "potentiality" to "action." If it is true that an experimental science requires a methodology, this is in no way sufficient, since its application is a function of the problems the subject formulates and, correlatively, of the range of possibilities she is capable of imagining. Now, the only problems the medieval methodologists were able to formulate about Aristotle's physics concerned certain points that were particularly questionable, rather than its value as a system: this undoubtedly required the collective movement of ideas, which did not begin until the Renaissance.

As for the relations between the "applied" logic of methodology and the "attributed" operations constituting causal explanations, these are fairly obvious: applied logic concerns reality and its description, which should present the facts and laws with as much precision as possible. This includes verifying explanatory hypotheses to see if they are borne out by the facts. In contrast, explanatory models have the effect of immersing reality within systems of compositions that are both possibility and regulated by necessity. Thus, reality gets simultaneously absorbed by possibility and necessity, but it remains real in that it retains the supreme power of decision over the experimental verification of the proposed hypotheses. It also gets enriched by this dual assimilation, which replaces pseudo-necessities by coherent causal necessity. In short, the subject is encompassed by reality as a physical-chemical organism, while her activities, which must necessarily adapt themselves at all levels to the environment—by using practical or scientific procedures or methodologies—are the origin of logico-mathematical structures. These encompass reality by integrating it with the other modalities, possibility and necessity.

III

The Historical Development of Geometry

I. THE EUCLIDIAN ELEMENTS

The history of mathematics does not begin with the Greeks. Taking the Greeks as a point of departure does not mean adhering to a tradition which has unjustifiably relegated—or at least minimized—the development of mathematics of the other peoples of Antiquity to a less important role, by regarding the "Greek miracle" as a discontinuity without precedent or parallel. But following the Greek tradition will enable us to work within a continuous history: in spite of many uncertainties as to the beginnings, it is possible to establish a process of which the successive stages can be followed step by step to modern times.

There is no doubt that, within Greek mathematics, geometry is the branch that has proved to have been so perfected that it has long been the paradigm of all science. Two thousand years after Euclid, it became the model for the construction of scientific theories for Newton, whose *Principia* were inspired by this model.

However, our aim is not to present a historical synthesis of Greek geometry, but to emphasise those characteristics that show the methodological bases and the epistemological framework within which the Greek geometers worked. To this end, we shall restrict our analyses to the four most important names associated with that period: Euclid, Archimedes, Appollonius, and Pappus.

Euclid is sufficiently well known to make it unnecessary to discuss in detail the historical significance of his *Elements*—doubtlessly the most important contribution of Antiquity to the methodology of science.

The value of his work remains regardless of the controversy over whether the *Elements* were written by Euclid as some scholars believe, or a school, as others claim. It is also outside of our concerns to discuss his predecessors, who, according to Proclus, had produced writings on the elements of geometry. That author speaks of Euclid as the one who "brought together the *Elements,* put in order many of the things discovered by Eudoxus, completed what had been begun by Theaetetus, and demonstrated more rigorously what before him had been shown in too loose a fashion."

The importance of Euclid's contribution, as formulated in his *Elements,* that is, to present the first axiomatization in the history of mathematics, was—(except for Archimedes) not appreciated until much later. It was not understood in all its depth until the turn of the twentieth century, on the basis of Hilbert's and Peano's work. We shall not insist here on analyzing this contribution. The importance of the axiomatic method is very well known and it is possible to find a clear presentation of it in any good manual dealing with the foundations of mathematics and logic. In addition, the axiomatization of a theory represents the final stage, the culmination, of its development; it is the systematic formulation of elements that have previously been elaborated, with the aim of clarifying their logical connections. Despite its usefulness, this method is far from being the only one of interest in epistemological analyses. Our analyses, as we stated above, concern specifically the process of construction itself, the genesis of successive structures, and the mechanisms mediating the transition from one stage to the next in the historical evolution of a science.

Euclid's *Elements* are of interest to our analyses because they represent, in a perfect fashion, the type of geometry that was to predominate during the entire period from Antiquity to modern times. These characteristics did not come to light until nineteenth century, at a time when a complete methodological revolution and a profound change in the conception of the significance of geometry took place. It is then that its characteristics and the limitations deriving from these were clearly evidenced. To understand this process, it is perhaps convenient to take a look at the development of geometry from the seventeenth century on and then go back to the Greeks. In this way, their accomplishments and the obstacles encountered will be given a sense in the light of later developments.

II. ANALYTIC GEOMETRY

After the Greeks, the first spectacular mutation was produced by analytic geometry. Even though preceded by Fermat (1601–1665), it was René Descartes with his famous *Discourse on Method and Rules for the Direction of Mind to find Truth in the Sciences* (1637) who was the principal instigator of this process. The third appendix of the *Discourse,* entitled "Geometry," is the milestone that marks the beginning of the modern period in the history of mathematics.

Descartes and Fermat were to replace the points in a plane by pairs of numbers and curves by equations. Thus, the study of the properties of curves was to be replaced by that of the algebraic properties of the corresponding equations. In this way, geometry was reduced to algebra. Descartes was himself quite aware of the importance of his work: The same year his book was published he wrote a letter to Mersenne, in which he states that his methods of analyzing nature and the properties of curves surpasses traditional geometry in the same way that Cicero's rhetoric surpasses a child's ABC.

Half a century after the *Discourse,* Newton published his *Principia* (1687). The differential calculus created by Newton and, independently, by Leibniz, was to give analytic geometry a scope that Descartes had not anticipated. Later still, James Bernouilli, Euler, and Lagrange were to complete the "reduction" of geometry to analytic methods.

From a historical perspective, it is possible to determine to what extent analytic geometry surpassed "traditional geometry"—what exactly this advance which Descartes referred to consisted in, and to what extent it remained tied to Greek tradition. On the other hand, this comparison allows us to better appreciate the characteristics and the limits of Greek geometry, as we remarked above. This evaluation was carried out in the beginning of the nineteenth century. Two French mathematicians, Poncelet (1788–1867) and Chasles (1793–1889) were the best interpreters of this process.

In the introduction to his famous *Treatise,*[1] Poncelet indicates clearly in what sense analytic geometry has gone beyond "ancient geometry":

> While analytic geometry offers, by its characteristic method, general and uniform means of proceeding to the solution of questions, which present themselves, to determine the properties of figures; while it arrives at results whose generality is, for all practical purposes out of bounds, the other proceeds by chance; its method depends entirely upon the sagacity of those who happen to use it and its results are almost always limited to the particular figure one happens to consider. Through the successive efforts of many geometers the particular truths uncovered have multiplied constantly, but rarely have methodology and general theory gained anything in the process.[2]

Poncelet sees the underlying causes of this state of affairs as follows:

> In ordinary geometry, which is often called synthetic, the principles are altogether different, the method is more cautious and more demanding; the figure is described, one never loses sight of it, one always reasons about magnitudes, real existing forms, and never does one derive consequences which cannot be represented, in imagination or perception, by tangible objects. As soon as these objects no longer have a positive, absolute existence, a physical existence, one stops. One is rigorous to the point of not accepting the consequences of an argument, which was established for a certain general disposition of the objects of a figure to apply to another equally general disposition of these objects, and which would seem to have all possible analogies with the first. In a word, in this restricted geometry, one is forced to repeat the whole series of primitive reasonings as soon as a line or a point have changed place from the left to the right of another, etc.[3]

For his part, Chasles, in his magnificent historical synthesis on the development of geometry[4] makes a similar comment:

> The geometry of Descartes, aside from its eminently universal character, distinguishes itself also from ancient geometry in yet another particular way which is worth pointing out: it is the fact that it establishes, by means of a single formula, entire families of curves; so that one cannot discover, in this way, any property of a curve without simultaneously coming to

know similar or analogous properties in an infinity of other lines. Until then, only certain particular properties of a few curves had been studied, each taken separately, and always with different methods, which did not establish any relationships between the different curves.[5]

III. PROJECTIVE GEOMETRY

Once analytic geometry had been difinitely constituted, a whole body of doctrines established itself, which led to a thorough revolution of mathematical thinking, of which Poncelet and Chasles are among the most important proponents. The quotations we have just presented are part of their reflections on the historical development of geometry. But they do not stop there; they go on to present their own interpretation of this process and draw conclusions for a new formulation of this science. This formulation was to predominate throughout most of the nineteenth century.

Rarely in the history of science can one so clearly follow the process by which such a fundamental change in the manner of thinking is produced. In this case, it is not necessary to provide a deeper analysis to laboriously find an adequate interpretation. It suffices to note the explanations given by the protagonists of the change concerning the origin of their ideas.

Having explicated the degree of generality of analytic geometry, and also the limits of "ancient geometry," Poncelet goes on to examine the causes of these differences:

Algebra uses abstract symbols. It represents absolute magnitudes by entities which have no value in themselves, and which let these magnitudes have all possible indetermination; then, it operates and reasons necessarily on the symbols of non existence as it does on quantities that are always absolute, always real. . . . The result must therefore also participate in this generality and be applicable to all possible cases, to all values given to the letters which enter into it.[6]

But Poncelet understands very well that this capacity for generalization is not limited to "abstract signs." In fact, he notes further on:

Now, one is led to all consequences, not only when one employs algebraic symbols and notations, but also each time one reasons about some undetermined magnitude. One abstracts from their numerical and absolute values; in a word, each time one reasons about undetermined magnitudes, i.e. one employs purely implicit reasoning.

The problem Poncelet is interested in is to find methods specific to geometry—those which do not involve algebra, but are independent of the figure so as to obtain the same degree of generality as analytic geometry. He formulates the question in the following terms:

Ancient geometry is bristled with figures. The reason is simple. Since general and abstract principles were absent, each question could only be treated in the concrete state, on the figure itself, which was the object of the question. Inspection of the figure was sufficient to discover the elements necessary for demonstration and solution of the problem. But, one was not without experiencing the disadvantages of this way of proceeding, difficulties in constructing certain figures and their complexity made understanding laborious and unpleasant. Particularly in questions involving the geometry of three dimensions, the figures sometimes became quite impossible, and this is where the disadvantages to which we are referring were the most noticeable.

This flaw of ancient geometry makes the advantage of analytical geometry, where this flaw is avoided in the happiest manner. Subsequently, one must have asked whether in pure, speculative geometry there was not also some way to reason without the constant assistance of figures, the disadvantage of which, even when they are easy to construct, is that they cause mental fatigue and slow down thinking.

The writings of Monge and the professorship of this illustrious master, whose methods have been preserved by one of his best known students, inheritant to his chair, have solved the question. They taught us that it is sufficient, now that the elements of science are formed and quite numerous, to introduce into our language and our geometrical conceptions these general principles and transformations analogous to those of Analysis, which will permit us to seize a truth in its pure and primitive state and in all its manifestations and therefore enable us to make fruitful deductions with ease. These make it possible to attain one's goal.[8]

Chasles follows a path parallel to that of Poncelet, and after having presented a historical study, which is still a classic in the history of geometry, he comes to identical conclusions:

In thinking about the procedures used in algebra and trying to determine the immense advantages they bring to geometry, is it not obvious that these advantages are partly due to the transformations one applies to the expressions one has first introduced? Transformations are the secret and the mechanism of true science, the constant object of the analyst. Was it then not natural to introduce into pure geometry analogous transformations that can be directly applied to the figures proposed and their properties?[9]

There is no need to give further quotations, since it is clear that Poncelet and Chasles were to incorporate systems of transformations as a fundamental method in geometry. In this way, they try to give to that science, but without recourse to algebra, the same degree of generality, of flexibility and productivity as analytic geometry had shown over the course of its development in the eighteenth century.

But is is also clear that these two geometricians were to introduce their new geometry on the basis of algebraic methods. Algebraic methods also inspired them in giving a "purely geometric" sense to the "imaginary" elements. We shall present an example of this, which is particularly relevant, historically. In the second chapter of the work to which we referred above, Poncelet studies the projective properties of figures. He analyzes the consequences of applying certain theorems that had been proved "under the conditions of general construction" to those cases where the points or straight lines of a figure are imaginary. Finally, he comes to the following conclusions:

a. Two or more hyperbolas, which are similar and similarly placed (*s* and *sp*), in a plane must have parallel asymptotes: consequently, they must have two common points and a common secant in the infinite.

b. Given any number of ellipses *s* and *sp,* then for any given direction there exists a system of "supplementary" hyperbolas that have parallel diameters of contact; since these are all *s* and *sp,* they must have a common secant in the infinite, which one can assume to be parallel to the direction given. Consequently, the given ellipses have a common secant in the infinite—that is, they have two imaginary points in common in the infinite.

c. Given that a parabola can be considered as an ellipse that is infinite in length, all parabolas *sp* meet at one point whose tangent is located in the infinite.

d. Two or more circumferences arbitrarily placed in a plane are obviously curves *s* and *sp* on this plane. Thus, it is possible to apply the above reasoning to them, which proves that they have an ideal secant in common in the infinite.

From this last proposition derives Poncelet's well known assertion:

> Circles placed arbitrarily in a plane are thus not entirely independent from each other, as one might believe on first thought; they also have two imaginary points in common in the infinite. Under this relation, they must possess certain properties at once belonging to their system as a whole and analogous to those which they possess when they have an ordinary secant in common.

The introduction of these ideas into projective geometry was to permit a remarkable generalization and simplification of several partial results. For example, the two fixed points that are situated in the infinite and through which pass all circles of a plane were called "cyclical points." These make it possible to apply to the circle the theorem according to which the number of points of intersection of two algebraic plane curves of degrees m and n is equal to the product mn (the circumference was apparently an exception, since its equation is of the second degree; the intersection of two circles should be a case where $m = n = 2$, whose product is thus, $mn = 4$).

On the basis of the cyclical points in infinity, Laguerre was able to give a definition of the angle formed by two straight lines. In general, the analytical expression of all the Euclidian metric properties presupposes the relation between the property in question and the infinite number of circular points at infinity or, alternatively, conics and quadrics, later introduced by Cayley under the name of "absolutes." It was Cayley's idea that the metric properties of figures are merely the projective properties in relation to the absolutes. Cayley's work, it should be noted, is based on his theory of "quantics" (homogeneous polynomials of two or more variables) and their invariants. His studies of projective geometry, thus, are derived from an algebraic perspective.

Cayley's ideas were further developed by Klein, who was able to give them a degree of generality sufficient to make possible a synthesis of all of geometry. Klein's central discovery was the projective nature of the non-Euclidean geometries as well as the demonstration that projective geometry is independent of the theory of parallels. On the basis of Cayley's conception of metrics, Klein showed clearly that, depending on the nature of the "absolute," all geometries can be obtained: when the absolute surface of the second degree is a real ellipsoid, a real elliptical paraboloid, or a real hyperboloid, one obtains the geometry of Bolyai-Lobatchevsky; when the surface is imaginary, one obtains the non-Euclidean geometry of Riemann; when it is a sphere, one obtains Euclidean geometry.

Klein's studies have opened the way to a new period of geometry: its incorporation within modern mathematics.

IV. THE ANTECEDENT OF THE NOTION OF TRANSFORMATION

The notion of transformation is thus at the basis of the new geometry which developed in the course of the nineteenth century. One may, then, ask whether this notion has any historical antecedents, and, in any case, why it took so many centuries before it was used and its role fully recognized. Similarly, in so far as this notion comes from analytic geometry (which in turn is based on the introduction of coordinates, and, through these, on the introduction of geometry to algebra and to infinitesimal calculus) we must ask about the historical antecedents of these methods. However, we must emphasize that our search for antecedents is not motivated by the goal to establish rank orders or to specify (as often done in historical research) the degree of originality of a particular author or school of thought. In a way, our research is opposed to this practice historians emply. Our goal is first of all to determine the factors that have prevented the development of certain ideas, conceived at a particular period but which then remained in an embryonic state, sometimes for several centuries. Our working hypothesis is that these facts are not due to chance and these periods of latency correspond, in general, to periods when other methods and concepts were being developed, without which those ideas could not be elaborated in depth.

1. The Greek Antecedents

Among Euclid's lost texts there were three books dealing with porisms. According to Pappus and Proclus, that work had even more depth than the *Elements*. Of this work only thirty propositions remain, which Pappus included in his *Mathematical Collection*. The meaning of the term "porism" has always intrigued geometers. In fact, this was a new notion added to that of "theorem" and of "problem." It seemed to be somewhat synonymous with these. Chasles, for this part, proposed a new interpretation for the meaning of porisms. He arrived at the following conclusions:

> If this book of Porisms had come down to us, it would have given us a long time ago a conception and development of elementary theories about the anharmonic ratio, homographic divisions and involution.[10]

Archimedes introduced a new concept, which, for the first time, made possible the quadrature of the space between a curve and straight lines. To this end, he conceptualized this space (which is frequently called "area of the curve" for short) as a limit toward which approximates the polygons included therein, when the number of sides are increased by bisection in such a way that the difference between them becomes smaller than any given quantity. Archimedes applied this method, which he calls the method of exhaustion, for the first time to the quadrature of the parabola. Undoubtedly, this method represents the germ of what eighteen centuries later would become integral calculus.

With his *Elements of Conics*, Apollonius introduced not only a remarkable number of new methods, but also a methodology and new concepts. In them one can discern the distant ancestor of the analytic geometry of the eighteenth century. Thus, it has often been said that Apollonius was the first to have made use of a system of coordinates in his geometrical demonstrations. A concerte example will allow us to appreciate to what extent Apollonius really used such a system and in what respect it remains far removed from those of Fermat and Descartes. The example, chosen for its clarity, is taken from Chasles's reconstruction of the method used by Apollonius. Chasles called it the "fundamental property of conics":

Let us imagine an oblique cone with a circular base: the straight line from the top of the cone to the center of the circular base is called the axis of the cone. The plane drawn by the axis perpendicularly to the plane of the base cuts the cone along two ridges and marks a diameter on the circle. The triangle that has as its base this diameter and for sides the two ridges is called the axial triangle. Apollonius supposes, to form his conic sections, the plane that cuts perpendicularly to the plane of the axial triangle. The points where this plane meets the two sides of this triangle are the *peaks* of the curve; and the straight line that connects the two points is a diameter. Apollonius calles this diameter *latus transversum* (transverse side). "Let us erect at one of the peak points of the curve a perpendicular to the plane of the axial triangle of a certain specified length, as we shall indicate; then, from the endpoint of this perpendicular draw a straight line to the other peak of the curve. Now, from any point on the diameter of the curve, let us erect, perpendicularly, an ordinate. The square of this ordinate, contained between the diameter and the curve, will be equal to the rectangle constructed on the part of the ordinate which is contained between the diameter and the straight line and to that constructed upon the part of the diameter contained between the first peak and the foot of the ordinate. Such is the original and characteristic property which Apollonius attributes to his conic sections. He then applies these, by means of transformations and sophisticated deductions, to derive all the other properties. It plays about the same role in his system as do the second degree equations with two variables in the analytic geometry of Descartes.[11]

Finally, concerning Pappus, we shall only indicate two aspects of his work, which may similarly be considered, in a way, as precursors of methods and concepts that were later to be used in projective geometry. The first is taken from his study of the quadratrix of Dinostrate. The second deals with his discovery of the anharmonic relation.

The mechanical generation of the quadratrix is well known: it is the intersection of the radius of a circle, which rotates about the center, and a diameter that is displaced in parallel direction with respect to itself. Pappus, then says that "This curve can be formed by surface positions or by the spiral of Archimedes." The first method:

Given a helix described on a right circular cylinder: from its points, let us draw perpendiculars onto the axis of the cylinder; these lines form the rampant helical surface; by one of these lines let us project a plane at a proper angle onto the plane at the base of the cylinder; this plane cuts the surface following a curve whose orthogonal projection onto the base of the cylinder is the quadratrix.

The second method:

Let us take an Archimedes spiral for a base of a right cylinder: then let us imagine a conic revolution having for axis the ridge of the cylinder taken from the origin of the spiral. This cone will cut the surface of the cylinder following a double curve; the perpendicular lines drawn from the different points of this curve onto the ridge of the cylinder in question will form the rampant helical surface; A plane drawn along a ridge of this surface, at a proper angle, will cut it following a curve whose orthogonal projection onto the plane of the spiral will be the required quadratrix.

The second contribution that we shall be concerned with is his discovery of the relation that Chasles christened relation in a harmonic function. As an invariant of the whole system of projective relations, it was to play an important role in nineteenth century geometry. In proposition 129 of Book VII of his "Collection Mathématique," Pappus described it as follows:

When four straight lines originate from a point, they form on a transversal drawn anywhere in their plane, four segments which are in a constant ration to each other whichever the transversal.[12]

That is, if *a, b, c, d,* are four points at which any transverse cuts the four lines straight lines, "the ratio ac/ad : bc/bd will be constant whatever the transversal."

The examples taken from the four Greek geometers we referred to show that the Greeks had, in an embryonic form, a certain idea of the use of coordinates (Apollonius) and of successive modifications of a figure tending toward a limit (Archimedes) as well as some use of the notion of transformation by projection (Euclid, Pappus).

From a historical-critical viewpoint and with regard to our epistemological objective, we now have to pose a number of questions. First, of all, can we propose that these geometers were precursors of Descartes, Newton, and the geometricians of the nineteenth century? Secondly, what is the reason that the methods sketched in the preceding examples were not further developed or even applied in the course of the following centuries? And why is it that these methods did not change the characteristics of Greek geometry as indicated above?

Before trying to answer these questions, we shall refer to some examples of "precursors" of nineteenth century geometry in order to clearly indicate the distinctive characteristics of one or the other of these.

2. Characteristic Examples of Sixteenth- and Seventeenth-Century Geometry

The credit for having first used a transformation explicitly to solve a geometrical problem may well go to Viète. In his works on spherical trigonometry, he used a transformation of spherical triangles, which he described in the following terms: "If from the three peaks of a spherical triangle figuring as poles, one described arcs of a large circle, the new triangle resulting from this will be reciprocal to the first triangle, both with respect to the angles and to the sides." Chasles, the author of this quotation, adds the following comment:

> Let us point out, however, that this *reciprocal* triangle is not exactly the *polar* or *supplementary* triangle, in which the sides are the supplements of the angles of the primitive triangle and the angles the supplements of the sides: two of the sides of Viete's triangle are equal to the angles of the proposed triangle, and the third side is equal to the supplement of the third angle. In this way, the perfect reciprocity of the two supplementary triangles, and thus that constant duality of the properties of spherical figures is absent in Viete's two triangles. But it is nevertheless worth mentioning his fruitful idea to transform the triangles in the manner indicated, in certain cases of trigonometry, since it represents a first step of the inventive mind and the first germ of the general methods of dualisation.

We owe to Snellius, in his *Treatise of Trigonometry* published in 1627, the correct application of the notion of supplementary triangle and the systematic use of the transformations of triangles that had been inaugurated by Viète. However, neither the abstract notion of *duality* nor the general concept of transformation were, as we shall see also in other domains, conceptualized as such at that time, even though they would seem to follow naturally and directly from Snellius's work.

After Snellius, it was Desargues (1595–1662) and Pascal (1623–1662) who made the next important step. Both took up again the study of cones, dear to the Greek geometers, in a much more general fashion. This made it possible to simplify and to largely extend the knowledge of the ancients about the properties of curves.

The simplicity of the ideas introduced by Desargues (the ideas which have served as a foundation for his theory) is such that one might consider them today as self-evident. Surprisingly, they were not applied before, and it took more than a century for them to become widely known and accepted as a natural method in geometry. The reason why calls for serious epistemological reflection.

Desargues and his brilliant student, Pascal, were working on two fundamental ideas. In the first place, Desargues envisioned completely arbitrary sections of a cone with a circular base. He conceived of the idea of extending to the three conic sections the properties of the circle, since these follow from the different ways it is possible to cut a cone with a circular base.

The work begun by Desargues and Pascal was continued by two of their students, La Hire (1640–1718) and Le Poivre. (The latter remained unknown until Chasles rediscovered him). Both make use of a method of transformations which set in correspondence the points and the lines of an arbitrary conical figure with the points and straight lines of a circle. In this way they discover the properties of alignment and of concurrence, which prefigure the descriptive geometry of Monge. Their work also includes anticipation of homological figures, the theory of which was not developed before Poncelet.

3. Why Transformations Are Late To Develop

The preceding historical considerations raise a central problem for epistemological analysis: Why were these transformations so late in appearing? We stressed the absence of transformations in Greek geometry, in spite of Euclid's porisms, Apollonius' coordinates, and the first modifications of figures by Archimedes and Pappus.

One might invoke complex reasons for the late appearance of the transformations, and these will be analyzed in detail in chapter 4. Here we shall merely note that the notion of transformation clearly appears only with the advent of algebra and analysis, disciplines that did not begin to develop until the sixteenth and seventeenth centuries.

We demonstrate in this chapter that the notion of transformation in geometry has an undeniable origin in analytic geometry. In this connection we would note: (1) Even though, as is usually stated in textbooks, analytic geometry constitutes the "application of algebra to geometry," it is inseparable from integral calculus. The latter, which was to develop in close interaction with geometry itself, took the entire eighteenth century to become consolidated. (2) It thus took until the eighteenth century for certain developments essential for algebra to appear, both for calculus and geometry, before the latter was ready for the foward jump initiated by Monge and "thematized" by Poncelet and Chasles.

To give an idea of the complexity of the process of interaction between the "traditional" problems of Euclidean geometry and the development of algebra and of calculus, it is perhaps sufficient to point out that it is only with Euler (well into the eighteenth century) that it became possible to show how the movements and the symmetries of figures are related to the problem of changing the axes of coordinates and how symmetry can be specified analytically. It is well known that in taking this route, Euler succeeded in showing that the displacement on a plane is a rotation or a translation followed by symmetry. This is the period of consolidation for the methods of calculus and of their purification by means of the problems posed in geometry (either directly or via mechanics). We have already shown, with the explicit statements by Poncelet and Chasles, how this interaction was to feed geometric thinking retroactively and thus, supported by algebra, take the big forward jump characterizing the nineteenth century.

V. THE FINAL STAGE: ALGEBRIZATION

The developments that took place in the first half of the nineteenth century continued to prevail until well into the twentieth century. They dominated the field of geometry until the ideas of Sophus Lie and Felix Klein, on the one hand, and of Riemann, on the other, made their appearance and their full implications were realized.

The period following the introduction of the concept of transformation such as conceived by Poncelet and by Chasles was a long one marked by a splendid apogee. Yet, the systematization of geometry which they achieved entailed a fundamental limitation: within it, the distinction between metric and projective properties cannot be strictly formulated. It is only with Lie and Klein, on the basis of the group of transformations and the corresponding invariants, that we are in possession of the tools necessary for introducing the precise distinctions between the different types of geometry. This time it was Felix Klein who was to give a masterly formulation to the new viewpoint. A new epoch was inaugurated. The passage from the period of projective transformations to that of group structures constitutes a particularly valuable lesson in epistemology.

The idea of transformation introduced during the preceding period had a clearly intuitive origin. The recourse to intuition had some obvious advantages, which we have already indicated, but also imposed certain limitations.

In each particular case a specific type of transformation was applied: this made it possible to study the properties of figures at a very high degree of generality, but the means by which to identify and express the *structure* of the set of transformations were missing. Group theory was to furnish the tools necessary for formulating the problems at another level, at which the structuring becomes manifest.

It is usually assumeed that Klein had become familiar with the theory of groups through C. Jordan's book (1870). However, his relations with Lie could very well have been at the origin of his first contacts with the works of Galois and Abel. We shall not attempt to elucidate the question to what extent Klein was original. What interests us here is the formulation he offered in his *Erlangen Program* (1872), which is, undoubtedly, the most clear and concise reformulation of geometry that has ever been realized.

Klein's conceptions had as a starting point the notion of the *group* of transformations of space. Now, as Dieudonne points out, "Klein's great originality has been to have conceived of the relation between a 'geometry' and its group by reversing the roles of these two entities, so that the group becomes the principal object, while the various spaces upon which it 'operates' show various aspects of the structure of group."[14]

Felix Klein clearly explains what he does this way:

There are transformations of space which do not alter in any way the geometrical properties of the figures. On the other hand, these properties are in fact independent of the spatial location of the figure in question, of its absolute size and finally also of the orientations of its parts. The spatial displacements, its transformations with similarity and those with symmetry thus do not change in any way the properties of the figures, no more than do the transformations composed with the preceding ones. We shall call the set of all these transformations the *principal group* of the transformations of space; *the geometrical properties of figures are not altered by the transformations of this principal group. The reciprocal is also true: the geometrical* properties are characterised by their relative invariance relative to the transformations of the principal *group*. In fact, if one considers for a moment, space as immobile, etc. as a fixed multiplicity, then each figure possesses its own individuality; those properties it has as an individual figure are the ones, the only ones that are properly geometric and which the transformations of the principal group do not alter.[15]

Klein, thus arrives at a fundamental reformulation of problems in geometry:

As a generalisation of geometry the following general problem arises: Given a multiplicity and a group of transformations of this multiplicity, to study the entities from the point of view of the properties that are not altered by the transformations of the group.

This definition he specifies further as follows:

Given a multiplicity and a group of transformations of this multiplicity: developing a theory of the invariants relative to the group.[16]

This completes the historical process that began with Monge and became explicated (thematized) by Poncelet and Chasles: that of the introduction into geometry of the notion of transformation. Klein was to realize he had made a great advance by going on from transformations to the sturctures that "explain" them. In fact, once it is recognized that the system of transformations that leaves certain geometrical properties invariant forms a group, one can replace the analysis of transformations themselves by that of the internal relationships of the groups. It is the relations between elements of a *given structure* that come to be of prime importance.

The transition from one stage to the other is equivalent to interrelating the various groups of transformations that characterize the different types of geometry, which come to be conceived as different subgroups of a global system that contains them all. This is what the "subordination" of the various geometries under a single group consists in; thus, each type of geometry becomes a special case of this unified group.

VI. CONCLUSIONS

1. The Notion of Transformation in Geometry

The notion of transformation first announces itself in Euclid's porisms and in Pappus' constructions, and would remain in an embryonic state for centuries. From the sixteenth century on, it would slowly make its way for three centuries. Occasionally, brilliant ideas would appear, but without achieving the status of a "universal method" for geometry.

In fact, the notion of transformation as such was not elaborated during that period. It was applied without geometers' being conscious either of its meaning or of its generality. There is nothing surprising about this. It can be explained with reference to a general rule that will frequently be demonstrated in this book. Abstract mathematical notions have, in many cases, first been used in an instrumental way, without giving rise to any reflection concerning their general significance or even any conscious awareness of the fact that they were being used. Such consciousness comes about only after a process that may be more or less long, at the end of which the particular notion used becomes an object of reflection, which then constitutes itself as a fundamental concept. This change from *usage,* or implicit application to consequent use, and *conceptualization* constitutes what has come to be known under the term "thematization."

The time that passed between the historical moment when transformations began to be used and that of their thematization is equivalent to the interval separating, say, the use of sets—distinguishing clearly between the elements of various sets—from the moment when the notion of set came to be thematized.

The delay between use and thematization does not come about by chance. Both our historical-critical analyses and the studies of psychological development converge in a surprising manner to show that this fact is of deep epistemological significance. In the present case, the basic reason for the delay in thematization emerges clearly from Poncelet and Chasles. Both authors recognize that it is through analytic geometry—that is, through algebra—that Monge developed his descriptive geometry. They similarly based themselves on analytic geometry and algebra to justify their methods and to develop their geometrical theories. Thus, it is, after all, the operational nature of algebraic transformation which constitutes the basis for its conceptualization in the field of geometry.

The time elapsed from Desargues and Pascal to Poncelet and Chasles is the period during which analytic geometry is consolidated. Its support consists of algebraic transformations. Geometry remains subsidiary to algebra. The transformations are applied by means of equations. Geometry appears only in the beginning of the process, with the formulation of the problem, and at the end, as a *translation of the result* of applying algebraic transformations.

It was to require a long period of uninterrupted work in algebra and infinitesimal calculus as well as in "translation of results into geometry" to finally come to a conceptualization of the very idea of geometrical transformations without *going through algebra or analysis.* It was necessary to "work" intensively with "negative segments" and "imaginary solutions" resorting constantly to algebra to arrive, for example, at the brilliant conclusion that all circumferences in a plane pass through two cyclical points in the same plane and to be able to demonstrate this by means of simple, purely geometrical transformations. As a principal "witness" of the difficulties of this process one might cite a geometrician as eminent as Carnot to whom the use of negative or complex quantities applied to the representation of geometrical "beings" appeared "absurd" and unintelligible. In his *Geometry of Position* he affirms categorically: "I shall show that this notion is completely false and that its acceptance would produce the wildest absurdities." This principle of negative quantities taken to be the opposite of positive quantities leads one imperceptibly into error."[17]

The conceptual distance that separates Euclid and Pappus, Viète and Snellius, and even Desargues and Pascal from nineteenth century geometricians is determined by the absence, in the former, of the essential operational tool: transformations. Until the nineteenth century, geometry had remained in about the same conceptual framework to which the Greeks had assigned it. The enormous progress made by analytical geometry did not succeed, by itself, in modifying this framework. The 185 years which passed between Descartes' *Geometry* and Poncelet's *Traité* (Treatise) were necessary for the development of the operational tool that was potentially capable of revolutionizing geometry.

Yet, during all this time, that instrument is applied only to extend knowledge about the properties of curves and figures without any essential change in perspective concerning these properties. Neither analytic geometry nor analysis has produced such change, even though they are the domains responsible for having developed the powerful methods which were to make this change possible. From the point of view of the conception of geometry, Descartes as well as Newton belong to the Greek tradition, even though their methods are far more advanced.

The preceding considerations furnish the elements for an attempt to elucidate the epistemological problems of historical development we formulated. This will permit a deeper understanding of the nature of the process at work; after this we can go back to history, following the connections that relate the most important stages in the history of geometry to one another, "explaining" each of these in their historical sequence.

2. The Three Types of Algebrization of Geometry

The application of algebra to geometry can be realized according to three perspectives each of which is appreciably different in conception and scope:

a. as a simple *algebraic translation* of the relations between the elements of a figure and a specific geometrical problem;

b. as an application of the notion *algebraic function* and of the *transformations* of functions;

c. as an application of the concept *algebraic structure* and of the relations between the elements of a given structure.

In case (a) we may assign, for example, a number to a segment, establishing a correspondence which is *fixed.* It is based on the disposition of the points of the segment and on the choice of a constant (unit of measurement). Thus, Apollonius obtains the fundamental property of conic sections from which all others can be derived. The relations which can be established with this method correspond strictly to the internal relations between the elements of a given figure.

Case (b) corresponds to the period inaugurated by analytic geometry and continued with projective geometry. The basic idea, as we noted above, is the notion of transformation. Unlike case (a), the algebraic representation does not correspond to a geometrical element of variable magnitude, but to a variable element within a system of possible transformations.

The change from case (b) to case (c) constitutes a new "relativization." Instead of establishing point-by-point correspondence between figures, it establishes correspondences between the elements of a given structure. More precisely, with the transfigural relations an inversion of the process is realized; no longer can one figure be transformed into another by verifying that certain conditions are met (maintain certain elements constant); we are dealing now with a structure operating on a set of elements. But to see clearly what all this means it is necessary to consider a few points of a different nature.

3. The Base Relations: Intra-, Inter and Transfigural Relations

The starting point of our analyses will be the conceptual framework developed by the Genevan School during their research in developmental psychology. The fruitfulness of this conceptual approach as applied to the history of science shows not only how the historical-critical and the psychogenetic studies converge—a thesis defended for many years by one of the present authors—but also how they may *interact* in the process of elaborating each of these topics. In this process, the notions derived from psychogenetic analysis have served as a guide in clarifying historical developments or even bringing to attention certain important aspects which a purely historical account would have entirely neglected. These notions will be presented in greater detail in the following chapters. We shall only briefly present them here in order to arrive at certain conclusions, which will lead us to an epistemological explanation of the evolution of geometry.

Geometry begins with Euclid—with a period during which the object of study is geometrical properties of figures and solids seen as *internal relations* between elements of figures and solids. No consideration is given to *space* as such, or, consequently, to transformations of these figures within a space that contains them all. We shall call this period *"intrafigural"*—an expression already used in developmental psychology to account for the development of geometrical concepts in the child.

The following period is characterized by efforts to find relationships between the figures. This manifests itself specifically in the search for transformations relating the figures according to various forms of correspondence. However, these transformations are not yet subordinated to structured sets. This is the period where projective geometry predominates. We shall call this period *"interfigural."*

Following next is a third period, which we call *"transfigural."* It is characterized by the predominance of structures. The work most characteristic of this period is the *Erlangen Program* of Felix Klein.

These three periods, clearly marked in the history of geometry, document an evolutionary process in the conceptualization of geometrical notions. Each period is marked, not by an increase of knowledge compared with the previous period, but by a total reinterpretation of the conceptual foundations. We have emphasised this throughout our analyses.

This evolutionary process reinforces the position long held by genetic epistemology showing, with many examples from developmental psychology, that cognitive development is never linear, but generally requires the reconstruction of what had been acquired during the preceding stages in order to advance to a higher level. It involves a reorganization of knowledge in the light of new information and a reinterpretation of basic concepts.

What is remarkable in this sequence intra-, inter- and trans- is the coexistence of three fundamental properties: First, the same sequence is found in all disciplines (at different developmental rates or different historical circumstances depending on the complexity of the domain to be explored) and with the same regularity in the succession of stages, as will be shown. Second, this process is not specific to scientific thought; the same *sequential order* as well as the same mechanisms have been found in the studies of the psychogenesis of concepts in the course of children's cognitive development (see the following chapter). Third, at each period the same process repeats itself according to certain phases inherent in the total process.

This last point requires a more detailed explanation, and we refer the reader to chapter 5, where we analyze again the deep epistemological significance we attribute to these three stages, giving evidence of the role they play in the development of the cognitive system at all levels.

IV

The Psychogenesis of Geometrical Structures

We have said that the psychogenesis of the notion of impetus is one that antedates modern physics. There it was possible to demonstrate a fairly detailed parallelism between the historical periods and the stages in psychological development. The historical evolution of geometry, however, goes far beyond anything that can be observed in the elementary stages. We shall nevertheless attempt to show that the processes of construction involved in this evolution are operative from the beginning, so that the same mode of functioning can be found at all stages in the transition from one to the next, even though the content becomes progressively richer and the structures stronger.

I. *Two specific traits characterize the mode in which space is constructed.* In the first place, there exists a space of objects (the physical) and a geometry of the subject (the mathematical). And even if the evolution of knowledge about the former depends naturally upon the instruments constructed by the latter, with a certain number of retroactions, there are nevertheless two distinct courses of development. Secondly, both of these kinds of space, the mathematical and the physical, in the course of their evolution, go through a period during which they are conceived as continuous totalities. Mathematical space incorporates the total set of figures. Each figure has certain general properties of the space and also—more importantly—constitutes one of its sectors. Physical space contains the total set of material objects. Each object is contained within this universal and permanent framework. In spite of this similarity, which seems evident, a divergence follows, which appears to be even an opposition, but which we need to interpret in order to discover its real significance. Following its "totalizing" phase, mathematical space undergoes a series of differentiations so that geometry comes to be more and more subordinated to algebra. At the same time, the notion of general properties of "space" is lost in favor of a multiplicity of structures, which can be coordinated, to be sure, but which no longer constitute a unique space. At the other extreme, so it seems, of this decline of "geometry" as a separate discipline, one observes a growing and fruitful geometrization of physics. It is as if the objects, which until then had been contained "in" space, now received from it their most important characteristics.

Given this dual process of evolution of physical and mathematical space, the later stages of which derive from a general need to explain the forms in terms of transformations, it becomes possible to recognize certain corresponding processes in the early stages of psychogenesis. From such observations, we can then specify the constructive mechanism and so identify it as the common mode of functioning which is also found in the transitions of one historical stage to the next, however rich the content and the level of complexity of structures.

The first stage in the psychogenesis of space is therefore that of intrafigural relations. If we ignore momentarily the domain of perception and speak only of representation, we observe even in the child's first drawings the distinction between open and closed figures, curvilinear and rectilinear ones, those which include right angles and those which do not, figures with different numbers of sides, etc. As for the contrast between intra- and interfigural properties, we can cite a good example. Two straight lines perpendicular to each other are easily reproduced as a single figure, but when it is necessary to make use of an external reference, children fail, for a long time, to trace horizontal and vertical lines correctly: thus, they draw a chimney on an inclined roof (even when copying it from a model) as perpendicular to the roof, that is, as slanting instead of vertical lines.

To these intrafigural relations we can add those which result from a comparison between the internal properties of two or more figures. This is quite different from interfigural relations seen as the position of figures within a surrounding space, which has to be structured since it presents the characteristics of a totality. For example, having discovered that the three angles of a particular triangle taken together produce a "half moon" (180°), young children succeed readily in predicting that this would be true of any kind of triangle whatever, and even that the angles of a square taken together would produce a "full moon" (360°). One can also consider certain geometrical locations as intrafigural; an example is the discovery that placing a series of objects at equal distances from a human figure will form a circle. In contrast, it is much more difficult for a subject to discover that the points of equal distance between two players are not limited to a single midpoint, but occupy the entire midline which can be extended in both directions as far as one might wish: in this case, where the line is perpendicular to the path relating the two players, its construction requires an organization of the plane and thus pertains to the interfigural.

But although it is true that the beginning of geometrical structuring is intrafigural and thus conforms to the historical order of succession, one of the present authors once noted that, unlike historical evolution but in agreement with the theoretical order, the first spatial forms children consider are topological in nature; only later are they able to handle Euclidian and projective figures. For example, only from four years on are children able to reproduce squares. Before that age, they represent these figures by closed curved figures (as opposed to crosses, etc.); furthermore, they are quite capable of drawing a small circle in the middle inside, outside, or at the border of this curved figure: "between the outside," as some of our young subjects expressed it. But though, at the intrafigural level, there is inversion of the historical sequence (while the interfigural and transfigural stages are later in development), we have to distinguish between the level of action where the first topological intuitions are found and the level of thematization with reasoning about the figures, where the interplay of morphisms about neighborhoods and topological spaces is far from primitive.

II. *There are three reasons for the change from intra- to interfigural relations in psychological development:* the necessity to render empty and filled spaces homogeneous; to coordinate directions or distances in two or three dimensions; and to localize mobile objects in the case of displacement.

Empty and filled spaces present a curious problem. For example, if one places two miniature trees 20 or 30 centimeters apart on a table, and asks a child whether the distance remains the same if one puts a partition about 2 to 3 centimeters thick between them, it is only from the age of seven, on the average, (the beginning of the concrete operational period) that children perceive that the distance between the trees remains unchanged. Before this age, children believe that the interval has decreased (although one might have expected just the opposite) because the space occupied by the partition does not have the same value of remoteness as the empty space. "If you leave a hole in the partition," some of our subjects told us spontaneously, "nothing changes; but if you fill in the hole, the distance gets shorter." This made it possible for us to use the device of the open or closed hole as an experimental control: this heterogeneity of empty and filled spaces amply demonstrates that initially subjects do not conceive of space in general as containing the objects or as a medium that relates one figure to another.

As for figures, let us examine the problems of distance or direction in the case where two or three dimensions are involved. For example, we present children with two identical sheets of paper and draw a point near one of the upper corners. Then we ask the children to reproduce this point in exactly the same place on the other paper. The youngest subjects content themselves naturally with a visual estimate. When somewhat later, they come to use measurement, they limit themselves, for quite a long time, to simply tracing a single oblique line from the point to the nearest corner: when they realize that their point is either too high or too far to one side, they start over, not understanding the insufficiency of this oblique measurement. Then, they may replace it by a new single measure, either horizontal or vertical. It is only with the onset of concrete operations (seven or eight years) that they come to understand spontaneously the necessity of two combined measurements to locate the point. This achievement means, analogous to those of Descartes and Fermat, identifying any point in a plane with a pair of numbers. Thus, we find here a behavior involving the use of Cartesian coordinates—of course only at the level of action and without thematization or even a reflective theory.

Another procedure to study the construction of a reference system consists in presenting subjects with a landscape or a miniature village (small buildings arranged on the table at various distances from each other) and to have subjects reproduce the pattern on another table. In this way, one obtains a complete hierarchy of strategies from simple alignments to correct structuring. The latter emerges later, when there are multiple elements as compared to a single point, as in the previous example.

To go on to directions, it is clear that tracing a horizontal or a vertical line requires that the subject establish interfigural references, rather than merely draw perpendicular lines, as discussed above. It is true that these are notions relative to the space of physical objects, but even after realizing that the water level of a bottle is horizontal and a lead wire suspended from a wall is vertical, it remains for the subject to reconstitute these lines, if there is no model to copy from. (Even copying is quite difficult for the youngest subjects.) For horizontal lines, subjects had to indicate, with a blue mark, the surface line of colored water in a transparent bottle that was tilted to various different degrees. Then we covered the bottle and presented subjects with 2 drawings depicting the same degree of tilt as the bottle. The subject was asked to mark the corresponding surface lines on these drawings. Now, until eight or nine years, children remain so fixated on intrafigural references that they draw a vertical line even when the bottle lies flat, because this conserves the initial situation where the water level was parallel to the bottom of the bottle. In other situations, subjects refer to the angles described by the bottle etc., but it never occurs to them to use external references, such as the surface of the table or that of a large supporting structure placed between the table and the bottle.

As for the vertical line, we had subjects draw a reproduction of a lead wire suspended from the top of a wall at either a vertical or an oblique angle. Again, when the wire hangs obliquely, children under eight or nine fail to complete the task.

III. A third set of factors leading the subject to consider interfigural relations have to do with the representation of displacement (much as the scholars of Merton College and Oresme were on the way to analytic geometry when they began working on geometrical descriptions of motion). Thus, in one of our experiments[2] we presented two small rods, *A* and *B*, with one pushing the other perpendicularly, either in the middle or at one or the other ends. In the latter case, a partial rotation is produced. The effect of the push of *A* against the center of *B* is, of course, understood quite early, since in that case, the displacement of *B* is only an extension of the movement of *A*. In contrast, the prediction of *B*'s rotation when *A* pushes against one of its ends becomes general only at about seven or eight years. These predictions, however, remain global; that is, they do not include the following two specific indications, which are not acquired until children are eleven or twelve. (1) There is as yet no composition of the translatory with the rotational movement with respect to the ends of the rods; (2) the displacements of *A* and *B* are seen as being related to each other, but not yet to their fixed support (cardboard or table surface). In either case, what is missing is the coordination of the two reference systems (internal and external). For, in order to become operative, the interfigural relations must be accompanied by interdependent transformations which thus become transfigural (the third level of this psychogenesis).

It is also noteworthy that the conservation of length is retarded in a situation where a ruler *A* is displaced parallel to a ruler *B* so that *A* extends slightly beyond *B*. In this situation, children under eight believe that *A* becomes longer and that the interval *aA—aB* (where *a* is the point of arrival of *A* and the endpoint of *B* on the same side) is longer than the interval *eB—eA* (where *e* is the opposite endpoint of *B* and *A* after *A*'s displacement). These judgments are contrary to what is perceptually observable. We thus find here nondifferentiation between displacement and lengthening, even though the interfigural relations of the positions "in" space are modified only by a very small movement.

This initial nonconservation of length leads us to examine in greater detail the boundaries between interfigural and transfigural relations, in other words, the level examined here and the level examined where the geometrical entities become the object of two structural transformations at the same time. Let us first recall that, historically, the construction of cartesian coordinates has immediately opened up the possibility of algebraic curves corresponding to polynomials and that of a geometrical solution of algebraic equations. Thus, we should not set the beginning of the period of transfigural relations before the moment when transformations come to play a role in the context of total algebraic-geometrical structures rather than that of simple figures—that is, at the time when the notion of "space" as a general container is replaced by multiple structures which can certainly be coordinated but are well differentiated. Hence, it is not surprising that in psychogenesis we reserve the stage corresponding to transfigural relations to a level of relatively complex structures like dual systems of coordinates, projective relations between several objects, etc., rather than having it begin as soon as there are transformations of simple figures.

The more elementary transformations raise the problem of what status they are to be given. The evidence just presented concerning conservation of length (where displacement plays a role precisely in terms of such a transformation) will help us to solve this problem. In fact, any change in the form of a figure is due to displacements of parts and any displacement may be expressed as interfigural relations, since what is involved is a comparison between initial and final positions with their respective reference systems. In this case, any figure subjected to changes of form constitutes a reference involving its initial and final form, while the parts displaced within this reference system represent, in their starting and final stages, figures or subfigures which are to be related in terms of interfigural relations. This is why when subjects focus their reasoning on internal interfigural relations with respect to such limited reference points, they are as yet only at the interfigural stage. They have not yet made use of the compositions and total structures which characterize the transfigural stage.

The simplest case of such changes in the form of figures is that where the parts are only displaced, without any other modification. One example is a square divided into four equal subsquares which are then reorganized as a rectangle; the subject must decide if the total surface has been conserved. In another example, four cubes first are joined into a larger cube, which in turn is placed on top of another such assembly of cubes, so as to result in a "tower." The subject is asked about conservation of volume. Now, even in simple cases such as these one finds that the youngest subjects, who are still at the intrafigural stage, deny the conservation of surface and volume (as they deny that of length in the case of the rulers). These invariants are not attained until the interfigural level, with its elementary reference systems. Note, too, that where the question concerns not simple geometrical figures but actual physical objects that have spatial properties related to their mass and that are indissociable from a spatiotemporal context, we find exactly the same reactions: when changing a clay ball into a sausage or pouring a liquid from a narrow breaker into a wider one, the parts are only displaced; but as long as subjects stay with intrafigural relations instead of reasoning in terms of displacements and their interfigural relations they refuse to accept the conservations of substance, weight, or volume. It is as if the displacements were accompanied by absolute gains or losses.[3]

From these kinds of reactions—either purely geometrical or spatiophysical—one can infer that the principal characteristic of displacement conceived in its interfigural form is what we call "commutativity."[4] This is the equivalence between what is added at the end of the displacement and what is subtracted at the start. Now, since they focus only on the point of arrival, young subjects interpret these changes as absolute gains, neglecting the subtraction at the starting point. In contrast, with the acquisition of commutability displacements become reduced to simple changes in position, hence the necessary compensation between additions and subtractions—in other words, the constitution of invariants in the face of modifications of figural forms. Furthermore, since the displacements are seen with reference to the total figure (that is, with reference to an external frame) rather than with reference to the parts, commutability concerns only this totality, and the change in global position is seen as having no effect on the distance between its various parts. In the case of a physical object, it is perceived as an undeformable or rigid solid. This notion, however obvious it may appear, is inaccessible to the youngest subjects before they reach the level of interfigural relations (instead of intrafigural relations) as well as of commutativity. In fact, commutativity requires that the undeformable solid should occupy, at the end of the displacement, a position that is equivalent to the one it began with. This is different when a figure is modified. In that case, commutativity is seen to hold for the individual parts displaced within the total figure. The latter is, then, the system of reference.

As far as related transformations on an isolatable figure are concerned, we have studied the modifications of a diamond-shaped figure by means of an apparatus called "Nuremburg scissors," which transforms, in continuous fashion, an equilateral diamond with lateral orientation into one with vertical orientation, while passing through an intermediate configuration of a tilted square. Young children expect to find that the figure gets larger or smaller in absolute fashion. They are not capable of conserving the form or the parallelisms. Somewhat later they gradually come to understand the correct transformations. These are then seen as an interplay of compensations between the lengthening of the short diagonal and the shortening of the long one, as well as between the equivalent changes in the two pairs of angles. All of this is relative to an external reference system (one in which the figure is not rotated). What we have here, then, is a series of interfigural correspondences with commutability in the wider sense of the compensations just described.

The projective transformations are understood and anticipated at the same level as the preceding conservations when they concern perspectives relative to a single object, while the relations between perspectives relative to several objects are more complex and involve transfigural modifications applied to a total structure. With respect to a single object, one can make it turn, move further away (creating perspective lines), or have the subjects themselves move. In all these cases, elementary projective relations involve different viewpoints—that is, the subject's positions and distance relative to the object and its different perspectives are also mastered at the same level, and it is remarkable to find again the same kind of development as in the cases just cited. To be sure, at first there is no prediction of change, since the object is seen as just an object, and the subject's viewpoint is not recognized as such. But as soon as subjects discover empirically, or begin to predict, a modification it is again considered as a kind of absolute change. The subjects have no comprehension of compensations. Thus, a disk or a watch presented horizontally on a flat surface or a pencil in oblique or "standing up" position is drawn as a half moon or as a half pencil since they cannot be seen in their totality at any single instant. In contrast, once the relations are understood, they obey the laws of compensation: when the object is rotated, the parts that become invisible are replaced by those that were previously invisible (back to front, or top and bottom, etc.). An object that moves away is drawn as getting smaller; when it approaches, as getting larger. Here again we find the construction of interfigural relations as a function of reference points, but in this case relative to different perspectives. The compensations involved correspond to an extended commutability. One might say as much about similarities, but here the invariance consists in proportions—that is, in relations of relations. This explains why such problems are more difficult and accessible only at the transfigural level, at eleven to twelve years.

IV. *It remains for us to describe the different forms of what we have been designating by the rather inclusive term compensation.* We use this concept to find the characteristics common to the various forms of commutability, which enable subjects to understand these compensations. Finally we have to contrast these with the compositions and operations which become possible at the transfigural level.

First, a distinction has to be made between compensations by inversion and compensations by reciprocity. Inversions involve subtractions from a given point, which are compensated for by additions at other points. In a simple displacement, for example, a mobile vacates the place it had occupied before its movement. Reciprocities involve reversals. Compensation results from, say, the reversal in the direction of a path; a modification due to an increasing function in one direction is compensated in the direction of a decrease in another. It is also important to distinguish an actual inversion of an object from one only relative to the subject's point of view. In the case of perspective changes, for example, there is inversion for the subject only; what has been visible now becomes invisible, and vice versa. The relations front–back or top–bottom, however, are reciprocities for the object only. As for reciprocities for the subject, we have to distinguish between those dealing with size and distance (as when an object gets smaller with increasing distance and, becomes larger as its distance diminishes) from those which concern the form (such as a circle which when perceived from an angle becomes an ellipse without a decrease in diameter and, inversely, returns to its circular shape after a correction in the positions).

Now, in the case of each of these interfigural modifications, it can be observed that cognitive progress follows the same law: at first the subject perceives the results but without understanding their relation to systematic transformations. This relationship is understood and mastered as soon as subjects are able to perceive possible compensations. The question then becomes, by what means are compensations discovered? In this respect, it seems that the process of commutability may play a mediating role, which may be generalized. Basically, it consists of relating the terminal state of a change to its initial state, or, more precisely, to subordinate the notion of state to its formation. But no matter how easy this seems to be, this change in direction causes a problem, since every action, whether carried out or observed, is oriented toward a goal, and the point of departure can easily be overlooked. Commutability involves first of all a directional inversion. This constitutes in itself the beginning of a cognitive construction. But, in addition, this change must be continuous if it is to attain the status of a transformation, rather than being concerned simply with the initial state in a purely static manner. (This is the only type of change very young children will recognize when they limit themselves to comparing a final state with an initial state and thus arrive at a judgment of nonconservation.) Thus, it is this process, requiring a comparison in the opposite direction, that leads to compensation and characterizes the second aspect of commutability: whatever changes or is lost in the course of the modification is, or may be, compensated for by what is gained and vice versa.

Briefly, commutability appears as the point of departure for operational reversible transformations. These first appear with modifications of an isolated figure at the level of interfigural relations. They achieve their final stage with a form of calculation when the results of two distinct systems are combined, as we shall show in respect to transfigural relations.

V. *When two systems are to be composed within a total structure, the elementary transformations based on commutability will no longer be sufficient.* They need to be combined. Examples are simultaneous displacements with respect to two kinds of referents and coordinates, the combination of rotatory and translatory movements, or the combination of vectors of different intensities and directions. These compositions, first directed by observations, then become anticipated and/or deduced, so that they take the form of calculations with reference to lengths. When these are expressed as numbers, they develop in the direction of algebra, as in the case of proportions.

A first example is that of relative movements:[5] a snail may advance on a board in either direction. In addition, the board may be displaced in either direction, with reference to an external fixed object that serves as a standard of comparison. Subjects are then asked to predict different positions of the animal relative to the fixed object, given different combinations of the two kinds of movement. One question is what has to be done to make the snail stay in the same place relative to the object even though both it and its board are in motion. Another example is that of three passengers in a train. Here subjects are asked to compare the times during which these passengers remain in a tunnel, if two of the passengers walk through the train in opposite directions, while the third remains seated. Now, simple as these problems may seem they are not solved before our final level (at eleven to twelve years), because they require composition between two systems of motion. But as soon as this coordination is acquired, it gives rise to a relativization of concepts which goes beyond the observable transformations; when asked which of three observers would see the greatest number of bicycle riders within a given period of time (if they follow each other at the rate of one per minute), one who remains seated by his door, another who walks in the same direction as the riders and a third who walks in the opposite direction, the subjects who find the correct solution often say things like: "It's as if the guy remained in the same place and the riders went faster." This is certainly not relativity of the level of young Einstein who asked a stupified conductor: "What is the name of the station that just stopped at our train?" Nevertheless, it is a transposition of correct relations, as far as the computation of equivalences is concerned.

The problem raised by these facts, and which we find again for all the others we observed, is why do these compositions of movements appear so late in development while the component movements are comprehended very early? In fact, we are dealing here only with translatory movements. The snail A, for example, can move in one direction with a motion $+A$ or in the other $-A$, and the board B with a motion $+B$ and $-B$: this yields four combinations $+A$ $+B$, $+A$ $-B$, etc. These divide into 12, depending on whether $A = B$, $A > B$ or $A < B$. Now, suppose these movements were successive, and one asked the following question: "If A covers distance $+A$ on the board but has to get to such and such a place (external object), what has to be done with the board?" There would not be the slightest difficulty for any of the 12 combinations. How can one explain, then, that if the translations $+A$ and $+B$ are simultaneous instead of remaining successive, the question becomes so complex that a child must be several years older before she is able to give the solution spontaneously? Now, from the point of view of spatial figures, the main difference is that the two movements to be composed in a simultaneous mode cannot result in a unified image, except by means of a motion picture. Each of the displacements corresponds to a single image. It can appear to be static or it can appear cinematic if one follows it with one's eyes or one's thoughts. In contrast, two simultaneous translatory motions are impossible to follow, even when they are moving in the same direction, because each is constantly modifying the other's positions while conserving its own characteristics. The notion of "transfigural" therefore takes on its etymological sense: that which no longer can be observed directly in a single figure and thus has to be computed. It is true that the interfigural relations already presuppose two figures, but these are only compared as states to be related by a transformation, which is itself of the figural type. In the case of transfigural transformations, on the other hand, the problem is to fuse two figures into one. This figure cannot be perceived and must be constructed by combining two transformations into a new product. As this perception requires an intermediate computation $(+A) + (+B)$, it is the elaboration of this computational mode that explains the late development of transfigural transformations. This is not the place to discuss the necessity to coordinate two systems of reference, which is obvious in the examples cited but remains implicit. More explicit manifestations of this will be discussed further on.

We shall now turn to a discussion of the coordination of translatory with rotatory motions. In these cases, the resulting figure may be represented as a single figure. However, this is in no way given by simple inspection of the components; it has to be constructed each time by means of inferential activities. A clear case is that of the cycloid—the result of composing the rotation of a circle in its perimeter and its translation, which can be observed directly in its center. Now, a cycloid cannot be perceived if one is looking at a wheel that is moving forward. Young children therefore represent it by a series of circles connected to each other by simple straight lines, somewhat later by contiguous circles. Interfigural comparisons may suggest global motion, but there is no way to understand the process of the transition from one rotation to the next, since they are merely juxtaposed. On the next level, subjects make connections but produce only epicycloids because of a predominance of rotations. It is again only at the age of eleven or twelve years that subjects become capable of constructing the cycloid.

A small board that is made to advance on top of a rotating cylinder may appear to be a far simpler composition. In this case the distance covered by the board is twice as long as that covered by the cylinder, since to the distance covered by the cylinder is added the movement it transmits to the board. Now, by looking at the figure described by these two mobiles, one can perceive quite well the advance the board has over the cylinder. But one cannot see that the board covers, in the same amount of time, the same distance that the rotating cylinder covers on the table. Again it is not the figure as such that provides comprehension, but rather a computation (no matter how simple this addition may be) of a transfigural nature.

Another composition we studied is that of a spiral. We had a pencil move in a straight line along a rotating cylinder. The first solutions consist again in circles related by line segments, then in oblique and parallel lines. Finally they are transformed into curved lines forming a spiral. In the case of the Archimedean screw[7] and the waves, we have not tried to solicit predictions of results on the basis of components, but instead asked for an analysis of the latter. Now, by glueing a piece of paper on the helicoidal tube which is subjected to an oblique rotation, children up to about seven years of age expect that the paper will rise to the top of this hollow screw where the water flows from one spiral turn to the next. After seven years of age, children no longer predict this effect, but they still do not see that the water level falls at each turn. It is only at around eleven to twelve years that the movements are described correctly. The same is true for the undulations described by a cord measuring between 3 and 4 meters or the small waves caused by a drop of colored liquid falling into a water basin. At first, children believe that a ribbon attached to the cord would make an advance motion and that the drop will cause waves to propagate over the entire surface up to the walls of the basin. Again it is only at about eleven to twelve years that subjects can dissociate the sinusoidal wave from the local vertical motions of objects describing movements independent of the wave.

These problems may seem complex, but they are all solved at the same level as the preceding ones, which demonstrates the general family resemblance between the modes of composition involved. In order to provide a better analysis of these compositions let us return to cases of the coordination of translations and rotations. Although apparently simpler, they clearly illustrate the difficulty of establishing necessary relations between systems of reference external and internal to the kinematic apparatus. Again we refer to the small rods mentioned in section III, one of which pushes the other perpendicularly, causing it to rotate.[8] These movements are correctly described in global and interfigural terms at about seven to eight years. But they also raise problems of a transfigural nature that are solved a great deal later; these problems arise when subjects are asked to describe in detail the translations and rotations of the extreme ends of the rods. Various experimental setups have been used to study this, of which the two main ones are presented below

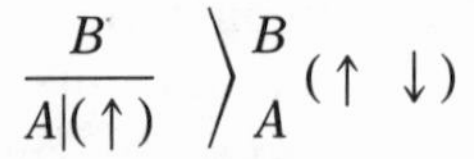

$$\frac{B}{A|(\uparrow)} \qquad \Big\rangle^{B}_{A} (\uparrow \downarrow)$$

One of these is simple and includes only one directional movement (pushing), while the other has articulations and allows for movements in both directions (pushing and pulling). The instructions are simply to describe the steps and the result of these actions and to indicate precisely the positions of the ends of the rods, particularly those for *B*. That is, the proximal position where the action of *A* takes its origin and the distal position (i.e., the free end). Now, in the case of the articulated rods, the distal end *B* descends when the proximal end *A* moves upward, and vice versa. These rotations can be followed mainly by means of the internal reference system (the relations between *A* and *B*), whereas the translations require a recourse to the external system (the positions with respect to the support). The result is a series of errors of coordination between these various movements (no need to describe these in detail).[9]

It is remarkable that until about eleven to twelve years of age, subjects do not even succeed—after having made erroneous predictions—in making a correct objective reading of the process underway. Now, since the movements of the game are all very simple, it seems clear that the main difficulty is that of composing into one the two systems of reference. Thus, the general problem is solved only at the usual level of transfigural constructions, which involve continual multiplications of relations.

To this coordination of reference systems there corresponds, in the domain of projections, the coordination of points of view with regard to a single set of several objects, for example, three cardboard mountains whose relative positions are to be predicted when looked at from four different perspectives around the table upon which they are displayed. In this case the relations to be combined are only those of left-right and in front-behind, with additional indications concerning the visibility of the objects. What seems to make the question easier is that children do not have to produce drawings, but have only to choose between a series of pictures for each of the four possible positions. Now, predictions turn out to be just as difficult as in the case of the two systems of coordinates and for the same reasons, which have to do with the necessity of simultaneous composition between the various relations involved—that is, of having to proceed by a logical computation.

As for compositions of vectors,[10] we find the constitution of explicit computations, but once again only at the final level and following a long series of trial-and-error behaviors. For instance, subjects who rightly predict that two equal forces will produce maximal performance when they act in concert while they cancel each other when they act in opposition will conclude from this a series of effects of decreasing magnitude with respect to the first situation—that is, as a function of increases in the difference of directions. Furthermore, the inverse problem of direction as a function of intensity is resolved at the same last level: that is, if the result of two equal forces, with a difference in direction of 90°, is located on their median, it will be closer to the greater force as a function of increasing difference in the intensity of the forces; this is, in fact, the law of the parallelogram, derived by simple logical or logical-geometrical computation.

With proportions, finally, we come to a computation that is both geometrical and numerical. Their spatial aspect can be uncovered with respect to similarities: constructing a triangle or a rectangle of exactly the same form but n times larger. The numerical aspect comes out, for instance, in problems where the speeds of two synchronous motions are to be compared perceptually (if a mobile covers a given distance in two minutes, where will it be at the end of eight minutes?). Now, elementary as they may appear, these kinds of proportions are generalized only at the transfigural level, since they constitute relations between relations and thus also require coordinations between distinct systems: c is to d as a is to b.

VI. In general *transfigural relations are thus quite different from interfigural ones.* This is true regardless of the multiplicity and diversity of their manifestations. In fact, interfigural relations place the separate figures within one single system that encompasses them. This is a homogeneous system, or isotrope: a system of coordinates or a structure characterized by one kind of transformation only (displacements, affinities, etc.). That transformation is generated from a general mechanism of commutability, which permits one to relate the successive states of a transformed figure to each other. Transfigural relations, by contrast, require that two actions be performed simultaneously, combining two distinct systems into one and joining together within a simultaneous set a number of relations that are easy to establish successively, but are not given as being associated in the initial figure. Transfigural relations thus characteristically substitute a computation for the description of figures, and even when the result of these compositions can be represented as a figure, this figure is new and must be constructed deductively before giving rise to a representation.

The mechanism of these successive spatial constructions is of a great simplicity even though it presents itself under three correlated aspects. The first of these is the change from elementary systems—intrafigural relations already constitute a system, a different one for each figure—to global systems, which can encompass them not only as subclasses, but also as containers of contained elements. This is a properly spatial relation. These global systems later change into coordinated systems, which are characterized no longer by relations of container to contained or by successive transformations, but rather by compositions uniting into a single action various connections that are all different from each other.

The second aspect of this constructive mechanism is indissociable from the preceding one: it is the change from figural relations to more abstract ones, constituting relations of relations of various degrees, such as proportions which represent equivalences between ratios (not to speak of second order relations as in proportions of proportions). Certainly, the beginning elaboration of complex relations can be found in the intrafigural domain, but the higher the rank of the system to be constructed the more the connections involved become removed from the purely spatial mode, which must combine with the logical-arithmetical computational mode, a prelude to a general algebra. Once orthogonal coordinates are present in a plane, a point can be represented by a pair of values that can be expressed as numbers. Double systems of coordinates require a logic of relations. We have seen how difficult they are to acquire up to eleven to twelve years of age.

The third characteristic of this common mechanism is the one that furnishes the motivating force for the two others and can thus account for them; it is the need to go beyond the facts to a comprehension of their reasons. Now, a spatial form cannot explain itself, and in itself it can only be the object of a figural intuition. In order to attain the reason for its properties, it is thus necessary first to subordinate these to quantitative laws in general, and then to conceive of the reason as being the result of transformations. Thus, a Euclidean axiom such as "subtracting two equal quantities from two equal quantities leaves two equal quantities as remainder" (we have verified, by modifying two surfaces in different ways, that this axiom is understood by children at the interfigural level only) concerns algebraic as well as geometrical quantities. On the other hand, the sum of the angles of a polygon when one increases the number of sides is not understood unless the transformations involved are related to a law of recurrence (without which the circle of 360° corresponding to the four angles of a square is simply made larger to obtain the result for a pentagon, etc.). In short, the integration of elementary structures within more powerful systems can be explained this way: to understand the forms one has to consider them as resulting from transformations, and to comprehend the latter one has to go beyond the geometrical toward a possible computation. One does so by subordinating the spatial quantities to the quantitative domain in general. Otherwise, the spatial entities remain purely figural and do not reach the level of internal necessity characteristic of logical-arithmetical structures.

Given this threefold mechanism of the construction of space in the course of psychogenesis, it is easy to see how it can give rise to completive and generalizing reconstructions at all stages in the history of geometry. Even considering the higher levels beginning with the Erlangen Program, one may consider each of Klein's fundamental groups as having given rise to intrastructural relations, so that their interrelations within a system of subgroups, where each is embedded within the other, constitute interstructural relations and that the rings and solids, in the following, considered by algebraic geometry, correspond to transstructural periods. In this case, the prefixes intra-, inter-, and trans- apply to structures of a higher rank, not to elementary figures.[11] But to speak of integration of simple systems within richer and more complex ones would be more than trivial if these generalizations common to all levels were not accompanied by the paradoxical phenomenon according to which the advances in geometry have led to algebrization, which eliminates geometry as such, and that superior knowledge of space has led to its suppression in terms of a general "container" and its replacement by more differentiated structures or fields. Now, these paradoxes, which characterize the true problem in the development of geometry, find their modest counterpart in the psychogenetic change from interfigural to transfigural relations, when the progressions in reflecting abstraction bring about a subordination of figurative observations to computation and multirelational compositions. The reason for this is, as was indicated above, that at all levels, figurative intuitions of spatial quantities become comprehensible only if they are subordinated to the general laws concerning quantity.

VII. *But there are still things to be explained.* Why, in synchrony with the algebraization of geometry—that is, with the progressive fading out of mathematical "space" as a unitary and general entity—does one observe a physicalization and even a dynamization of the geometry of objects? This occurs in correlation with the progressive elimination of a unitary physical "space" as a general "container" of all phenomena. Now, in this respect and in spite of the apparent audacity of this attempt (if it is not simply candor), we believe that the psychogenetic point of view may contribute to a partial solution of the problem. The reasons for this are simple and obvious. First, we have to recall that the acquisition of knowledge is due neither to the subject's becoming conscious of her own actions nor to a simple reading of the properties of objects. Instead, knowledge comes from interactions between the subject and the object. These interactions are indissociable at first. Later, there is differentiation into interiorization (in which logical-mathematical relations are based on internal coordinates of actions) and exteriorization (oriented toward physical objectivity).

Now, the spatial connections occupy a central position in these initial undifferentiated interactions, since the subject's actions are deployed in space and the most elementary property of objects is to have positions in space. Thus, space constitutes, from the most primitive forms on, the fundamental instrument of the encounter between the subject's activity and the characteristics of the object. It is thus natural that the progressive (and very slow) differentiation between logical-mathematical constructions and physical knowledge deprives it of its priviledged and exclusive status and subjects it to transformations, which finally lead to two complementary and bipolar requirements—those of algebraization and of physicalization.

1. The lack of differentiation of the spatial and the logical domains is greatest at the sensorimotor levels, since knowledge is here reduced to simple knowhow, not yet conceptualized. Its only instruments are movements and perceptions—that is, actions which are carried out materially. Nevertheless, sensorimotor intelligence is structured in action schemata and their coordinations. But these schemata, while presenting the logical properties of identity, equivalence, etc., which are inherent in any generalization, evidently remain spatial because of lack of conceptual thought or representation. The same is true of their coordinations, since the relations of order, embedding, and correspondence—which characterize them—concern material actions carried out in space.

When, between two and seven years of age, this undifferentiated logical spatial schematism is completed by the formation of conceptual systems made possible by the symbolic or semiotic function, the two aspects of developing cognition, the logico-arithmetical and the geometrical, beginning to differentiate. But it is interesting to observe for how long a time concepts and numbers conserve certain spatial characteristics. For instance, subjects will correctly label a row of n tokens, but when the tokens are spread out so as to make a longer row but without any addition of tokens, subjects believe that the row now contains a different number of tokens ($>n$). Similarily, the first classifications are partly based on relations of similarity and difference, but require, in addition, that the collection of elements have a certain spatial configuration (figural collection), where intension and extension are undifferentiated.

At the level of concrete operations (between seven-to-eight and ten-to-eleven years), the differentiation of logico-arithmetical and geometrical structures seems to be completed, giving rise to two isomorphous but distinct systems of operations: the logical operations based on similarities and differences and the infralogical ones based on neighborhoods and continuities. But in as much as the former remain "concrete," i.e., their use requires the manipulation of objects, these operations still include spatial displacements, and hence an important residue of spatial aspects. It is only at the hypothetical-deductive level (from eleven to twelve years) that the logico-arithmetical structures are entirely freed from geometrical elements. It is precisely at this level that the "transfigural" phase becomes established in the subject's geometry—i.e., the beginnings of algebraization, however modest.

2. Three essential consequences derive from these facts of psychological development.

The first is that given the common origin of the logico-arithmetical and geometrical structures at the level of sensorimotor acts, space remains at all levels a necessary (but not sufficient) mediator between the subject and the objects. This is why even in theoretical physics, the structure of groups, for example, which is used to explain phenomena, conserves a spatial dimension, since wherever there is motion, real or virtual, there is some "field" with its extension.

The second consequence, also deriving from this common origin, is the isomorphism existing between the space of objects in its various manifestations, and certain aspects of subjective geometry (which at first is dominated by the former, but then becomes more and more enriched). In fact, while the subject proceeds only by successive approximations in her deductive reconstructions of the dynamics of objects, never being sure she has reached the limit, the spatial characteristics of experience are transparent to the mind to the extent that they remain geometrical.

The third consequence results from the fact that given this common origin due to the initial interactions of subject and object, knowledge then differentiates in two opposite directions: the internal coordination of action, which is the source of logico-arithmetical structures, and the links existing between objects, which physical concepts attempt to capture. Following this, space, which initially constituted the place and medium common to all sensorimotor knowledge, is transcended on two sides. It is transcended internally by means of logico-arithmetical or algebraic structures, and externally in terms of the dynamics of objects. The geometries of the subject and of objects are then no longer partially fused and begin to become dissociated in symmetric fashion, one in the sense of algebraization, the other in the direction of physicalization and dynamization. In either case, space ceases to play the role of a general container and becomes diversified into particular mathematical structures and various kinds of physical "fields." Thus, on either side, it gets attached to separate subfields of knowledge, which originated from the two conjoined movements developing toward the internal mechanics of subjective activity and toward the dynamics of objects. Of course, this does not mean that spatial structures lose their function as mediating functions between subject and objects, but that they are not the only ones to exercise this function, since knowledge is no longer centered uniquely on external objects.

VIII. CONCLUSIONS

The common mechanisms we have tried to discover between psychogenesis and the history of science were to be concerned with the process of regulating the sequence of developmental levels irrespective of their absolute rank. It now remains to ascertain the significance of the sequence intra-, inter-, and transfigural first in its general and then in its specifically spatial characteristics.

1. From a general point of view, the sequence, which we find in all domains and at all levels, is the expression of the conditions imposed upon all cognitive acquisition by the laws of assimilation and equilibration. In approaching a new domain, the subject must first assimilate the data to her own schemata (of action or conceptual). Whether the data are objects, figures, relations etc., their

analysis always involves an elementary form of equilibration between their assimilation to subjective schemata and the accommodation the latter have to undergo as a function of the objectively given properties.

This explains the "intra" characteristic of the first cognitive acquisitions. But the new schemata thus constructed cannot remain in isolation: sooner or later the assimilatory process will lead to reciprocal assimilations, and the requirements of equilibration impose upon the schemata or subsystems, thus interrelated, more or less stable forms of coordination and transformations. This explains the "inter" type, characteristic of the second level. But a third form of equilibrium will have to develop from this, since the multiplication of subsystems endangers the unity of the whole.

Thus, the necessary differentiations will be counterbalanced by trends toward integration. The equilibrium necessary between differentiations and integration can thus only lead to systems of interaction so that the differentiations can be generated instead of being imposed. This is the only way to maintain harmony and avoid internal perturbations or conflicts between them. This explains the global formative structures characterizing the "trans" level of development. These triads are much more flexible in principle than are the theses, antitheses, and syntheses of classical dialectics, even though they too are based on the role of disequilibria and re-equilibrations towards higher levels. Obviously, they are only artificially delimited phases of a continuous process. The structures attained at the "trans" level in turn give rise to new "intra" analyses, which lead to other interanalyses, followed by the production of superstructures of the type "trans," and so forth *ad infinitum.* If this mechanism is so general and constantly repeatable, it is then quite natural that it can be found in the transitions from one level to the next both in psychological and historical development, irrespective of the absolute height of the levels considered.

2. But the interpretation just proposed concerns for the most part only external aspects of equilibration. What remains is to indicate the internal aspects—that is, the increases in the coefficients of necessity characteristic of the successive forms of truth, in particular the geometrical ones. Now, it is obvious that logico-

mathematical necessity has an endogenous source with maximal values for closed systems of transformations constructed by the subject. In contrast, the data from external exogenous sources remain in a state of fact and have a *minimum* of intrinsic necessity. From this point of view, the change from intra to inter to transfigural modes thus corresponds to systematic increases in necessity due to this fundamental process of progressive replacement of initial exogenous by endogenous constructions. Here we have, thus, another aspect of the equilibration process analyzed above; it is in close association with the first.

Now, the case of geometry is special with respect to the relations between the exogenous nature of its beginnings and endogenous transformations. On the one hand, space is both a property of objects and a product of possible constructions by the subject; this tends to favor exogenous contributions, because a subject, in the process of constructing figures, etc., generally believes in finding pre-existing realities rather than applying her transformations to freely selected, abstract entities. On the other hand, being generalizations, the subjective endogenous constructions sooner or later have to taken on algebraic forms, while the interactions between the algebraic instruments to be constructed and the spatial forms to be projected to assure this union become more and more complex. This has the effect of extending the interfigural period and of retarding the transfigural one by time intervals that were sometimes difficult to understand at the time but are all the more interesting for retrospective epistemology.

3. One might thus imagine the following succession of our three levels, considered from the point of view of exogenous or endogenous forms of truth. The initial form is, naturally, figural realism. Figures are considered on the one hand as existing independent of the subject (in the physical or ideational world), and on the other, as resulting from themselves and not from any kind of construction emanating either from nature or from the subject. The latter can certainly manipulate the figure and introduce certain modifications in the form of dissociations of parts or of additions. But all this is done with the aim of gauging the intrinsic properties of the figure rather than of constructing a system of transformations of which the figure would seem to be the re-

sult (an exception to this are the conic sections which, however, attain the interfigural level at least partially). Given this perennial static nature of figures impinging on the subject from the outside as ready made "beings," the notion of transformation as a source of new constructions has no meaning within the realistic position. In fact, this mode of knowledge, determined by such initial conception of truth, in terms of simple *adequatio* can only be intrafigural, precisely because it involuntarily, but frequently also intentionally, maintains its exogenous character: the subject, being ignorant of her own operations or refusing to attribute to these any kind of capacity of constructing intrinsically true findings, will relegate them to the role of receptors commanded by the permanent "beings" given by the outside world.

With the arrival of interfigural organizations, however, endogenous constructions begin to play a formative role, in the sense that the geometrical "beings" are no longer simply imposed from the outside, but become an integral part of a set of transformations and relations, even though they owe their existence to the subject's operational instruments (now considered as a source of objective knowledge, as in the case of algebra). The two principal acquisitions resulting from this subordination of exogenous data to beginning endogenous constructions both result from comparisons between intrafigural relations, but accompanied by generalizations performed by the subject and which transcend the initial limits. The first of these acquisitions proceeds from a generalization of the relation of enclosure, given from the beginning in the perception of any closed figure. This generalization is easy to trace in the domains of psychological development. For example, when young children displace a figure or an object (as in conservation experiments), they usually pay no attention to the space left empty in A when the object passes from A to B, while at the operational interfigural level, they take into account both the empty and the occupied places. This is the point of departure for a new and more abstract form of enclosures, which will lead to an inferred and not simply a figural notion of space as a general enclosing container. From there it is only a short step to the construction of elementary coordinates.

The second acquisition then consists in discovering, within this general framework, the covariations between related figures. We have observed many qualitative examples of this in the child.[12] But it is obvious that their translation in the terms of algebraic functions and interfigural curves raises quite different problems, since this involves higher levels of scientific thought. It is, then, all the more interesting to find, even at these levels, for the forms specific to space, very complex relations between the exogenous and endogenous sources of knowledge.

The historical question can, in fact, be formulated in the following paradoxical manner: on the one hand, Greek geometry, not having algebra, remains of the intrafigural kind, subordinated to exogenous sources. Hence the absence of any transformation in spite of Euclid's "porisms" (local transformations oriented toward figural results), the partial coordinations of Apollonius and Archimedes, and Pappus' modifications of figures, which are all special cases without any methodological generalizations.

The subordination of space to algebra goes back to Viète and has remained entirely local (transformations in spherical trigonometry). On the other hand, in spite of the systematic attempts to establish correspondences between algebra and geometry inaugurated by Descartes's work, there were 185 years between it and Poncelet's *Traité:* The problem is, then, how to explain why the interfigural period lasted that long—that is, why almost two centuries went by before transformations were first used in geometry, while algebra, the science of transformations *par excellence,* had been applied to geometry from the seventeenth century on. Finally, with Lie and Klein (with some promising starts made by Chasles and Poncelet, but limited to projective geometry) the primacy of transformations finally gets established, subordinating *all* the geometries to algebraic systems.

It can be seen to what extent this problem is analogous to those raised in psychological development, since in fact it is due to the many difficulties encountered in the course of a laborious process of equilibration between endogenous and exogenous sources of knowledge. Algebra is in itself a system of forms which generate their own contents, while intrafigural geometry is in itself nothing more than a system of forms modeled obligatorily after a content already given, seen as being linked to objects. Algebra enters sure-footed into the domain of intrinsic necessities, while intrafigural geometry is "subordinated" only to its contents, which does not constitute autonomous necessity, but only a constraint of a legal nature. This is not at all the same thing. First of all, while in the domain of algebra the subject can feel free to construct the transformations that seem useful, faced with the idea of geometrical transformations judged possible, she can only ask herself whether or not she is entitled to apply them in the light of the "reality" imposed by the data (Carnot, it should be recalled, described as "absurd" the transformations used by Chasles and Poncelet). In sum, the long interfigural period can in no way be reduced to a history of collaboration between two kinds of instruments that can readily be intercoordinated. Rather it is characterized by difficult adjustments (given the duality of space as both a subjective and an objective reality) between two heterogeneous types of truth, which require novel instruments in order to be reconciled.

Let us recall in this respect that infinitesimal calculus needed the entire eighteenth century to develop its foundations and to overcome the difficulties raised by the notion of limit. Let us remember that it was only with Euler that the symmetry of figures could be translated into analytic terms, by means of a change of the rectangular axes. With J. Bernoulli and d'Alembert, new relationships are established between the composition of movements and the existence of instantaneous axes of rotation. Cramer's "determinants" play a role in the "analysis of curved lines." No need to comment on the obstacles he had to overcome before he could assign a geometric role to imaginary numbers. In brief, unlike other domains—where the "inter" period is one of regular progression in the replacement of exogenous factors by endogenous ones—the interfigural period is marked by a series of constructions to be carried out in order to adjust algebra to space, and vice versa.

Once these local conflicts are overcome, the general line of transfigural development consists then in subordinating all intra- and interfigural acquisitions to systems of sets of transformations, which generate the figures, or differentiated subsystems, instead of having to submit to their resistance; hence the primacy and final victory of endogenous factors elaborating the structures which no longer consist in "figures" (such as a group) but integrate all realizable constructions within total systems. Nevertheless, such situations never have anything final about them, because these structures, once elaborated and hence intrinsically necessary, can in turn be treated as data, as a kind of "pseudo-exogenous" reality, thus becoming a potential object of new intra-type analyses (cf. in morphisms coming after the operational structures thus leading to new constructions of the inter and trans variety).

Now, this interpretation of the three levels, in terms of exogenous, exo-endogenous (if one might use that term), and finally more and more endogenous truths enables us to give an acceptable sense to our efforts to uncover the common transition mechanisms between one level and the next (let us insist on this point), both in psychological development and the history of the sciences. Comparing the levels with the same labels in the two domains, one cannot help experiencing concern, since although there is convergence with respect to the sequence of levels, as far as contents are concerned, considerable differences remain. However, in speaking of replacement of exogenous knowledge reconstructions, we may have found the most general law of the evolution of knowledge: even before she has language, the sensorimotor infant empirically (exogenously) assimilates objects to her action schemata, and at the early age of 15 to 18 months, she can already coordinate these schemata in ways which Poincaré had long ago compared to groups of displacement. Since we are not interested in the contents of the developmental levels, but in their mode of construction, it does not seem to be more far fetched to compare the mechanisms involved in the sequence of stages in history to that found in psychological development than it is to look for common evolutionary mechanisms at vastly different levels of zoological phyla. Furthermore, since human intelligence is a special case of "behavior," and since in biology the evolution of behaviors remains full of mysteries, we are pleased to end this discussion by reminding the reader that even at very elementary levels there are certain clear cases of a mechanism called "genetic assimilation" or "phenocopy," etc., which consist in replacing an exogenous variation by an endogenous reconstruction. One of the present authors has already carried out a biological study of some of these cases where he attempted to show a certain kinship with what the psychogenesis of knowledge has taught us.[13] If this kind of generalization can be justified, it should have certain implications for the epistemology of science.

V

Algebra

In the preceding chapters we have shown the existence of common mechanisms to explain the evolution of the conceptual framework in the history of geometry and the psychogenesis of geometrical notions. In the historical domain we make use of an interpretation resulting from the psychogenetic research of the last decades. The development of knowledge occurs not by the continuous accumulation of new facts (accompanied by the refutation of concepts and hypotheses which turn out to be unfruitful or false), but in stages, which represent characteristic cognitive levels such that at each stage there is a reorganization of previously acquired knowledge.

Specifically, we have seen that the historical development of geometry, just like the psychological development of geometrical structures, is characterised by three periods, which we have called intra-, inter-, and transfigural. In close correspondence with these three periods, we can also distinguish three main stages and several substages in the evolution of algebra and—as we shall see in the following chapter—in the evolution of the logico-arithmetical relationships in the child. We shall call them, respectively, intra, inter, and trans*operational* stages.

The *intraoperational stages* are characterized by intraoperational relations, which manifest themselves in forms that can be isolated. These include, as their name indicates, internal elements, which, however, do not combine with each other; they do not include any transformations, which presuppose the existence of invariants.

The *interoperational stages* are characterized by correspondences and transformations among the forms that can be isolated at previous levels, along with the invariants required by such transformations.

The *transoperational stages* are characterized by the evolution of structures whose internal relationships correspond to interoperational transformations.

This analogy of developmental stages between geometry and algebra, on the one hand, and the history of science and psychogenesis on the other is remarkable. What is involved is not a simple classification of stages. In fact, the three notions constitute three different, but associated forms of organizing knowledge. We perceive these notions of intra, inter and trans as the most important and the most constructive of the mechanisms we have been able to find in our search for common mechanisms between history and psychogenesis. We shall analyze the epistemological significance of these mechanisms in the concluding chapter, after discussing their scope in the next two chapters.

The identification of stages turns out to be more difficult in algebra than in geometry (or in physics, as we shall see further on). The process of algebraization of mathematics constitutes in itself a transoperational stage in so far as the algebrized branches are concerned, yet several stages in the development of algebra cannot be interpreted without taking their interactions with analysis and topology into consideration. The detailed study of these interactions would in itself be a long-term project involving the total reconstruction of the history of mathematics according to our epistemological perspective. We had planned to carry out such a study, but had to abandon the project—at least within the limits of the present volume.

By limiting our original intentions, we proceeded in a somewhat arbitrary manner to select certain topics for inclusion here. Even though the analyses of each of these topics do not have the value of a demonstration, they are nevertheless useful as illustrations of the kind of hypotheses to which we have been led in studying the common mechanisms in history and psychogenesis.

We shall begin with an interpretation of the origins of algebra. This section has two aims. On the one hand, we shall expose both the difficulties inherent in an epistemological analysis of the historical process that leads to the constitution of a new branch of science and the virtual impossibility of using the accounts of this process as presented in current texts of the history of science. On the other hand, we hope to uncover certain mechanisms of the cognitive process (complementary to the intra, inter, and trans kind of mechanisms), which are particularly difficult to uncover in a scientific discipline that is already established.

The other topics are chosen from a particular branch of algebra—algebraic geometry—which exhibits a particularly interesting interplay of levels. Algebra and geometry, analysis and topology here interact in a very precise manner, lending significance to the evolutionary process which led to Grothendik's monumental achievement. We shall limit ourselves to sketching some of the critical moments of this process to substantiate our conclusions.

1. The Origins of Algebra

1. There are many historians of mathematics who trace the origins of algebra to various nations of Antiquity: the Assyrians, Babylonians, the Egyptians. Others, with more critical judgment, locate this origin at the school of Alexandria. Diophantus is generally considered to have been the first to have formulated arithmetical problems in symbolic terms and, to solve these problems, to have introduced "indeterminate values" represented not by numbers but by letters to express unknown quantities in equations.

This kind of historical interpretation has always appeared as fairly unsatisfactory. It is clear that the difficulties the Greeks encountered in resolving their numerous geometrical problems can be explained only by the absence of a science of algebra, which would have enabled them to formulate these problems in terms of operations. However, it is difficult to explain the almost total eclipse of a science, which did not resurface until the middle of the sixteenth century.

In the historical interpretation to which we were referring, Viète appears as a man of the Renaissance in the strict sense of the term. His "return to the Greek sources" is said to have enabled him to go back to Diophantus' science and to perfect it, so that it became the point of departure for the algebra of the modern era. In an interpretation of this type, however, the role the Arabs played remained relatively unclear aside from the introduction of a more adequate notation of arithmetical operations, that of the idea of zero as a number (a concept imported from India) and that of the generalized use of letters to represent "indeterminate" quantities. In this context, Viète's work appears like that of an erudite scholar, a work of systematization more than that of a creative genius and innovator of science.

The kind of panorama we described has undergone some thorough modifications since the publication of Jacob Klein's *Die griechische Logistik und the Entstehung der Algebra* (1934), whose greatest impact made itself felt only after the publication of its English version in 1968.[1]

Klein introduces a fundamental reinterpretation of the works of Diophantus and of Viète, based on a scholarly analysis of Greek thinking and of the sense of the "new science" which developed in the sixteenth and seventeenth centuries.

Klein's careful study has enabled us to situate the origins of algebra within the general schema of the mechanisms which we have found in the development of other branches of mathematics and physics as well as in the more recent stages of algebra itself. His chapter on the difference between ancient and modern conceptualizations furnishes the elements upon which our own interpretation can be firmly based. Thus, we shall begin with a concise exposition of Klein's position, with which we are in entire agreement, although it seems to us to require a complementary epistemological explanation.

The point of division between Diophantus and Viète hinges on the differentiation one has to establish in the way these two men use mathematical symbols. Once we have established that point it will influence all the historical interpretations we identify ours with. The algebraic character commonly attributed to Diophantus' *Arithmetic* is based on the use he makes of various signs and abbreviations referring specifically to unknown quantities in equations. These have usually been interpreted as constituting an algebraic symbolism. Now, it is obvious that the mere use of letters to represent numbers of geometrical entities does not render the treatment of a problem symbolic. Euclid, and before him Archytas, used such representations. In this connection, Klein cites the opinion expressed by Tannery: "A letter may well replace some arbitrary number . . ., but only where the number is supposed to be placed; it does not symbolize its value and is not used as such in operations." And Klein goes on: "Aristotle also used such mathematical letters, for example in his *Physics* and in *De Caelo,* and he even introduced them into his logical and ethical investigations. But this type of letter is never a "symbol" in the sense that what the symbol stands for is in itself a "general object."

It is only from the sixteenth century on that the use of letters took on a symbolic character. When attributing the invention of algebra to Diophantus (or to his predecessors whose work he is assumed to have compiled), one implicitly or explicitly takes a position on the question of the symbolic character of his methods to resolve mathematical problems.

The fact that Diophantus speaks of "general" problems and of a "general" solution may support the classical interpretation. Nevertheless, Klein's reinterpretation casts doubt on the symbolic character—in the algebraic sense of the term—one might attribute to these expressions. In this connection, he introduced a fundamental distinction between "the generality of a method" and the "generality of the object of investigation." Let us quote Klein:

> ancient mathematics is characterized precisely by a tension between method and object. The objects in question (geometric figures and curves, their relations, proportions of commensurable and incommensurable geometric magnitudes, numbers, ratios) give the inquiry its direction, for they are both its point of departure and its end. The way they determine the method of inquiry is shown especially in the case of "existence" proofs, i.e., demonstrations that the "being" of a certain object is possible because devoid of self-contradiction (cf. 104). The problem of the "general" applicability of a method is therefore for the ancients the problem of the "generality" (καθόλου) of the mathematical objects themselves and this problem they can solve only on the basis of an ontology of mathematical objects. In contrast to this, modern mathematics, and thereby also the modern interpretation of ancient mathematics, turns its attention first and last to method as such. It determines its objects by reflecting on the way in which these objects become accessible through a general method (pp. 122ff)

2. Viète took up a methodology characteristic of Greek thought, but he gave it a scope and a depth which enable him to reorganize Diophantus' work on a very different level. In our view, it is to Klein's credit that he has shown, with scholarly care, what this reorganization consisted of and why Viète must be considered the true founder of algebra.

The *Introduction to the Analytic Art*[2] includes a presentation of what Viète calls "a certain way of seeking the truth," which he considered characteristic of mathematics and whose discovery he attributed to Plato.

The term "analysis" given to this form of investigation originated, according to Viète, with Théon, whose definition he cites as follows: "taking the thing sought as granted and proceeding by means of what follows to the conclusion and comprehension of the thing sought." In the current sense of the term, "synthesis" is a process which begins with "taking the thing that is granted and proceeding by means of what follows to the conclusion and comprehension of the thing sought."[3]

Klein mentions that Pappus had offered a clearer explanation of the dual process of analysis and synthesis.

As for analysis, Viète reintroduces another distinction made by the Greeks, of two genres: *zetetic* (theoretical analysis) and *poristic* (analysis by problems). Then he adds a third kind, *rhetic* (exegetic) analysis.

there is a zetetic art by which is found the equation or proportion between the magnitude that is being sought and those that are given, a poristic art by which from the equation or proportion the truth of the theorem set up is investigated, and an exegetic art by which from the equation set up or the proportion there is produced the magnitude itself which is being sought.[4]

After this there follows a paragraph that is crucial in the interpretation of the originality of Viète's position:

And thus, the whole threefold analytical art, claiming for itself this office, may be defined as the science of right finding in mathematics. Now what truly pertains to the zetetic art is established by the art of logic through syllogisms and enthymemes, the foundations of which are those very stipulations (symbola) by which equations and proportions are arrived at, which stipulations must be derived from common notions as well as from theorems that are demonstrated by the power of analysis itself. In the zetetic art, however, the form of proceeding is peculiar to the art itself, inasmuch as the zetetic art does not employ its logic on numbers—which was the tediousness of the ancient analysts—but uses its logic through a logistic which in a new way has to do with species. This logistic is much more successful and powerful than the numerical one for comparing magnitudes with one another in equations, once the law of homogeneity has been established; and hence there has been set up for that purpose a series or ladder, hallowed by custom, of magnitudes ascending or descending by their own nature from genus to genus, by which ladder the degrees and genera of magnitudes in equations may be designated and distinguished.[5]

Essential to Viète's formulation is that the term "magnitude" is used in its most general sense. The magnitude in question is either a particular number or a specific measurable geometrical magnitude. Let us quote from Klein:

Hence arises the double name of this third kind of analysis whose business it is to effect the computation of "arithmetical" as well as the construction of "geometric" magnitudes starting from canonically ordered equations; it is called rhetic with respect to the numbers to which it leads and which can be expressed by the ordinary number names of our language; it is called exegetic in respect to the geometric magnitudes which it makes directly available to sight. (p. 167)

As Klein notes, we have here the convergence of two independent lines of pursuit: the geometrical analysis of Pappus and the arithmetic methods of Diophantus. Viète's "new" algebra was both geometric and arithmetic. To achieve this, it was necessary to attain a higher level of generalization than the "ancients" had achieved.

In the beginning of chapter 4, Viète introduces another illuminating distinction: "numerical reckoning *(logistice numerosa)* operates with numbers; the reckoning by species *(logistice speciosa)* operates with species or forms of things, as, for example, with the letters of the alphabet."[6] Here the key word is "species" *(espèces)*. Klein's meticulous study refutes, in our view conclusively, the currently accepted interpretations, even though these are just as authoritative as that given by Cantor. Klein's own interpretation shows what the essence is of Viète's formulation. This point merits a full citation:

The species are in themselves symbolic formations—namely formations whose merely potential objectivity is understood as an actual objectivity. They are, therefore, comprehensible only within the language of symbolic formalism which is fully enunciated first in Viète as alone capable of representing the "finding of finding," namely "zetetic." Therewith the most important tool of mathematical natural science, the "formula," first becomes possible (cf. Note 239), but, above all, a new way of "understanding," inaccessible to ancient episteme is thus opened up.

When we look back to the Pythagorean and Platonic concept of the eidos of an arithmos as that which first makes the unified being of each number possible (cf. p. 55f) and compare it with the concept of the species developed above, we may say that the ontological independence of the eidos, having taken a detour through the instrumental use made of it by Diophantus (cf. p. 143f. and 170), here finally arrives at its symbolic realization. This heralds a general conceptual transformation which extends over the whole of modern science.[7]

Viète's crucial distinction, which enabled him to constitute algebra as a new discipline, lies in his jump from the concept of "arithmos" to that of general symbols. *Arithmos* refers directly to things or units while the symbols (letters) he used make direct reference to the property of "being a number," a property that is common to all numbers; arithmos refers indirectly to the entities or units whose "numerosity" is represented by a number. In other words, the letters designate the concept of "number in general." Klein, dealing with Viète's conceptions, notes: "The letter sign designates the intentional object of a "second intention" (intentio secunda) namely of a concept which itself directly intends another *concept* and not a *being*."

Even though Eudoxus, Aristotle—and particularly Proclus—had proclaimed a *divina ars*, which is none other than the general theory of proportions, capable of encompassing all mathematical knowledge in its entirety, Viète goes a step further in penetrating more deeply into the concept of transformation. Here we differ with Klein, who fails to emphasise this point, which we consider crucial. The fifth chapter of the *Isagog*, entitled "the laws of zetetics," seems to us to contain an absolutely essential aspect of Viète's formulation and, in our view, its most important epistemological foundation. Klein relegates this chapter to a single reference (pg. 250).

The chapter in question contains the laws of transformation of equations. According to Viète, there are three such transformations: the *antithesis* (the transfer of a term or member of one equation to another); *hypobibasm* (the reduction of the degree of the equation by dividing the two sides by the *species* (common to all members of the equation); and *parabolism* (the division of the coefficients of an equation by a given quantity). It is no accident that Viète concludes this chapter with the following observation:

> Diophantus, in the books he wrote on arithmetics, applied zetetics more subtly than did anyone else. But he presented it as if it was established by numbers only and not also by species (of which he did make use, however), so that his subtleties and skill should be all the more admired, since the things that appear subtle and abstruse to one who uses numerical logistics (logistes numerosus) are totally familiar and immediately obvious to one who uses species logistics (logistes speciosus)[8]

It seems to us that it is in that chapter that one finds the true basis of the reasoning which consists in disregarding numbers and working instead with "species."

This interpretation of the underlying reason for Viète's "quantum jump" in advancing to a higher level of generalization to establish his new algebra had been explicated by the author who directly continued this line of thought: Descartes.

II. SOLVING ALGEBRAIC EQUATIONS

After Viète, the study of algebra was limited to equations until the mid-nineteenth century. The method for solving quadratic equations was discovered before Viète by the Hindus. Indeed, even the Babylonians had solutions to quadratics. Equations of the third and fourth degree were not solved before the end of the sixteenth century: the quarrel between Tartaglia and Cardan about the discovery of the formula with which cubic equations can be solved is well known.

There were, for a long time, many efforts to find a formula to solve equations of a degree higher than the fourth. Now, the only real success of this period is in the solution of systems of linear equations. One also finds, during this period, algebraic solutions to certain specific problems proposed by geometry or mechanics.

Nevertheless, each problem requires its own method of solution, a particular procedure. This indicates that we are dealing here with a period corresponding to the one we have characterized as *intraoperational.*

Thus, it is not surprising that there was very little progress during the seventeenth century and the first half of the eighteenth. Mathematicians during this long period of time centered their attention on the new instrument created by Leibniz and Newton: the infinitesimal calculus. This instrument, in the hands of mathematicians of the caliber of Euler, Lagrange, and Gauss, was to lead albegra—during the second half of the eighteenth century—to a new level of development. It was then that mathematicians succeeded in formulating, algebraically, problems of great generality—among them the fundamental theorem of algebra. To accomplish this, they took advantage of the properties of continuous functions and their transformations taken from infinitesimal calculus. Following our general definition, this period corresponds to an interoperational stage. For a long time, transformations were to dominate algebra until the emergence of the first algebraic structure, the group. This achievement was to lead the theory of algebraic equations toward its transoperational stage.

The central figure in the transition from the intraoperational to the interoperational stage is Lagrange. The empirical efforts to solve equations of various degrees (characteristic of the intraoperational phase) gave way with Lagrange to a question of more general scope: what exactly was the nature of the solutions given to equations of the third and fourth degrees and what was the reason for their success? Lagrange believed it was possible to derive in this way ideas that would help him to approach equations of higher degrees. He succeeded in showing that all the methods consist in introducing functions which "reduce" the original equation. The problem posed in this way comes down to finding the relation between the solutions of the reduced equations and those of the original ones.

Having arrived at that point, he was to use another very fruitful idea, which already contains the seeds of ideas which later led to the theory of groups: the number of different values a polynomial will take when its variables are permuted in all possible ways.

Lagrange analyzed certain functions of the roots of an equation and showed that the number of values a function y of roots $x_1, x_2 \ldots x_n$ can take when the x_j are permuted in all possible ways is a divisor of n. For example, for an equation of the fourth degree whose roots are x_1, x_2, x_3, x_4, the function

$$y = x_1x_2 + x_3x_4$$

takes only three different values when the roots are permuted in all 24 possible ways. Furthermore, Lagrange demonstrated that the number of different values determines the degree of the reduced equation which permits the solution of the equation in question.

Ruffini was to take up Lagrange's ideas again in trying to demonstrate the impossibility of finding a solution by radicals to a general equation of the fifth degree. Even though his demonstration remained incomplete, the conceptual framework in which he worked places him in an exceptional position during this interoperational period of algebra, very near the following stage which Galois inaugurated.

Ruffini defined the permutations of variables in a given function and classified them in genres. It is possible to change the terminology he used and to translate it into the terms of the theory of groups of substitution. Now, this is not simply a matter of change in terminology. According to Ruffini, the permutations are related to the values of the roots. In his view, the class of permutations that does not change the value of a function lacks structure. He conceived of the transformation implicated in the change from one permutation to another, but failed to conceive of the structure in which the transformation is in turn implicated.

For his part, Cauchy considered functions in a more general way, as "functions of n quantities"; but these quantities are not considered as the roots of equations. They are nothing but letters representing quantities.

Cauchy called permutations the order of the letters. The transition from a permutation A_1 to another A_2 he called *substitution* and denoted it as:

$$\begin{pmatrix} A_1 \\ A_2 \end{pmatrix}$$

Then he was to define the multiplication of substitutions and the identical substitution, which led to the introduction of the inverse substitution:

$$\begin{pmatrix} A_1 \\ A_2 \end{pmatrix} - 1 = \begin{pmatrix} A_2 \\ A_1 \end{pmatrix}$$

On the basis of these definitions he demonstrated a number of theorems, which we can consider as immediate antecedents of the general theorems about the groups of substitutions. However, he neither explained nor thematized the structure of the group as such.

Gauss' *Disquisitiones Arithmeticae* occupy a unique place at the end of this period. We refer, in particular, to the fifth section entitled: *Forms and indeterminate equations of the second degree.*[9] Although Gauss studied the quadratic forms only in relation to the solution of indeterminate quadratic equations, his careful analysis of the properties of binary and ternary quadratic forms was to become his main theme. Not only did he classify the forms as such by defining their "orders" and "types," but he also succeeded, for the first time in the history of mathematics, in "composing" these forms with one another, that is in defining operations between these forms.

Let us first consider some of the general definitions before showing the kinds of problems studied by Gauss. The fifth section of his *Disquisitiones arithmeticae* begins as follows:

In this section we shall treat particularly of functions in two unknowns x, y of the form $ax^2 + 2\,bxy + cy^2$ where a, b, c are given integers. We will call these functions *forms of the second degree* or simply *forms.* On this investigation depends the solution of the famous problem of finding all the solutions of any indeterminate equation of the second degree in two unknowns as long as these unknown values are integers or rational numbers. This problem has already been solved in all generality by Lagrange, and many things pertaining to the nature of the forms were proved by this great geometer. Still other properties were discovered in part first by Euler and later by Fermat. However, a careful inquiry into the nature of the forms revealed so many new results that we decided it would be worthwhile to review the whole argument from the beginning—the more so because what these men have discovered is scattered in so many different places that few scholars are aware of them, further because the method we will use is almost entirely our own, and finally because the new things we add could not be understood without a new exposition of their discoveries. We have no doubt that many remarkable results still lie hidden and are a challenge to the talents of others. In the proper place we will report on the history of important truths.

Further on in the same text, he gave a more precise definition of the term form:

We represent the form $ax^2 + 2bxy + cy^2$ by the symbol (a, b, c) when we are not concerned with the unknowns x, y. Thus this expression will denote in an indefinite manner a sum of three terms: the first a product of a given number a by the square of some indeterminate number; the second a doubled product of the number b by this indeterminate number times another indeterminate number; the third a product of the number c by the square of this second indeterminate number. For example, (1, 0, 2) indicates the sum of a square and double a square.

Additional essential definitions are the following:

We shall say that a given number is *represented* by a given form if we can find integral values of the indeterminate form so that its value is equal to the given number.

We shall see in what follows that the properties of the form (a, b, c) depend in a special way on the nature or this number $b^2 - ac$. We shall call it the *determinant* of this form.

If the form F with unknown x, y can be transformed into another form F' with unknown x', y' by substitutions like

$$x = \alpha x' + \beta y', y = \delta x' + \delta y'$$

where α, β, γ, δ are integers, we will say that the former *implies* the latter or that the latter is *contained in the former.* He shows that:

$$b'^2 - a'c' = (b^2 - ac)(\alpha\delta - \beta\gamma)^2.$$

And if further the form F' could be transformed by a similar substitution into the form F, i.e. if F' is contained in F and F is contained in F', the determinants of the forms will be equal, and $(\alpha\delta - \beta\gamma)^2 = 1$. In this case we call the forms equivalent. We will call the substitution $x = \alpha x' + \beta y'$, $y = \delta x' + \delta y'$ a *proper transformation* if $\alpha\delta - \beta\gamma$ is a positive number, *improper* if $\alpha\delta - \beta\gamma$ is negative. And we will say that the form F' is contained *properly* or *improperly* in the form F, if F can be transformed into the form F' by a proper or improper transformation. So if the forms F, F' are equivalent, we have $(\alpha\delta - \beta\gamma)^2 = 1$, and if the transformation is proper $\alpha\delta - \beta\gamma = +1$; if it is improper $\alpha\delta - \beta\gamma = -1$. If many transformations are all proper or all improper, we will say they are *similar*. A proper and an improper form will be called *dissimilar*.

On the basis of these definitions, Gauss formulated and solved the following kinds of problems:

1. Given any two forms with the same determinant, see if they are equivalent or not, if they are properly so or improperly, or both at the same time. If they have unequal determinants, see if one does not include the other, properly, improperly, or both at once. Finally, find all transformations, both proper and improper, of one to the other.

2. Given any form whatsoever, see if a given number can be represented by it. Find all the representations of a given number M by a given form F.

The great care with which Gauss studied these questions and the rigor with which he proceeded to analyze particular cases gave him the right to say:

In the preceding discussion everything that pertains to equivalence, to the discovery of all transformations of forms, and to the representation of given numbers by given forms has been satisfactorily explained.

Somewhat further on, Gauss comes to one of the most original points of his work, which he introduced as follows: "We will go on to another very important subject, the composition of forms."

Gauss' definition of the "composition of forms" constitutes the first operation ever introduced into a nonnumerical domain and whose properties cannot be directly deduced from operations on numbers. Gauss thus arrives at important theorems such as the following:

If g is a primitive form of the same order and genus as f, and g' is a primitive form of the same order and genus as f': then the form composed of g and g' will be of the same genus as the form composed of f and f'.

This statement is followed by the following comment:

From this we can easily understand what we mean when we speak of a genus composed of two (or even several) other genera.

From the preceding we can conclude that the "forms" studied by Gauss and the properties he demonstrated with so much care can be translated into the language of today as that of the theory of groups. In fact, a quadratic form, with a law of composition like that defined by Gauss, is an Abelian group having as a unit element the class that Gauss called "principal class." Analogous "translations" can be established on the basis of results obtained by Ruffini and Cauchy. Most historians are baffled by the fact that no answer can be given to the following question: why is it that these authors, having come so close to the concepts of the theory of groups could not take that additional "small step" necessary for its constitution? From our point of view, the question can be answered as follows: that "step" only seems small. In fact, Gauss—together with Lagrange, Ruffini, Cauchy and a few others—are the last representatives of the interoperational period in the development of algebra and specifically in the history of the theory of algebraic equations. His method consisted essentially in transforming the functions and finding the relations that remain stable. The properties he derived are simply invariants in systems of transformations.

The kind of development that we have found once again in the history of science, as well as in psychological development, indicates that there is a long road to travel before a system of transformations can be related to a total structure in which these transformations constitute intrinsic variations. This is, in fact, the step from interoperational to transoperational connections.

In psychological development, the transoperational stage is reached—as we shall see in the following chapter—when children become capable of carrying out operations on operations. For example, a child will discover how to derive all the possible permutations of *n* elements at the time when she becomes capable of systematizing the different permutations—that is, of introducing an ordering into the permutations she carries out. In that case, the set of permutations derives from a seriation of seriations.

It is not without interest to note that Galois introduced the notion of group on the basis of the action of "grouping." The following definitions constitute the point of departure:

> The permutation from which one starts in order to indicate the substitutions is quite arbitrary, when one deals with functions. For there is no reason why, in a function of several letters, one letter should occupy a particular rank rather than another.

However, since it is hardly possible to imagine a substitution without a permutation, we frequently make use in language of the term permutation and consider substitutions only as transitional phenomena between one permutation and another one.

When we wish to group substitutions we shall have them all originate from the same permutation.

A little further on, he states explicitly:

A system of permutations such as, etc. is called a group. We shall represent this set by G.

Here at the very source of the notion of structure formulated for the first time in the history of algebra, we find Ariane's thread which helps us understand the transition from the inter to trans operational stage. Just as in psychogenesis, this change presupposes a change from operations on elements to operations on operations.

III. FIELDS

The final stage in the study of algebraic equations was reached in the second half of the nineteenth century, a period during which one of the great historical "leaps" can be observed. We have seen examples of this characteristic process of scientific thinking in the historical evolution of geometry; we shall see it again with the development of physics. Fundamental to the process is a *reinterpretation* of variables—one of the most important mechanisms in mathematics. (In our chapter on physics we shall see that it is only a special case of a much more general mechanism). It will be useful to stop to analyze it in some detail in relation with the study of algebraic equations.

We have already cited Gauss' work on quadratic forms as a decisive step. The "composition of forms" he introduced constitutes the first "operation" in which numbers are not directly involved. Clearly, however, each of these forms was necessarily a relation in which coefficients and the variables both represent numbers. The next step was to show that the properties of the polynomic functions—and consequently of algebraic equations, which had until then been the object of study of algebra—do not depend on the coefficients and the variables being numbers. Once again progress was made by modifying the fundamental questions in the study of polynomials of different degrees. The question was no longer, what kind of number determines the properties or the zero values of polynomials? It became instead, what are the properties of numbers that are important to take into account? The answer was surprising: it had become clear that these properties were very general and did not constitute the characteristics of numbers as such. Many sets of elements possess the same properties as numbers do. The properties common to such classes of elements, thus, do not define a specific domain of mathematical objects, but a structure common to many domains. It was Dedekind who studied the structure in question, to which he gave the name "field."

In studying polynomials and algebraic equations one can thus abstract from numbers as such and consider as "coefficients" only those classes of elements that fulfill the following conditions:

I. Between the elements of a set two laws of internal composition hold

$$a, b \rightarrow a + b;\ a, b \rightarrow ab$$

These are called addition and multiplication, respectively.

II. These two laws of composition form a group (excluding, in the case of multiplication, the neutral element of addition, commonly called zero).

III. These two operations satisfy the following conditions, called distributivity:

$$a(b + c) = ab + ac$$
$$(b + c)a + ba + ca$$

It is easy to show that many "entities" that are not whole, rational, real, or complex numbers (as had been used until then) satisfy these rules and consequently constitute fields. Let us take as an example the following polynomials

$$a_0 + a_i x \ldots + a_n x^n$$

where the a_i are any rational numbers. We shall simply write:

$p = (a_0, a_1 \dots a_n)$

We can then define the sum and multiplication in the following way:

Given that

$P = (a_0, a_1, \dots a_n)$ and $Q = (b_0, b_1, \dots b_n)$

then,

$$P + Q = (a_0 + b_0, a_1 + b_1, \dots a_n + b_n)$$
$$PQ = (a_0b_0, a_0b_1 + a_1b_0, \dots, a_0b_n + a_1b_{n-1} + \dots + a_nb_0)$$

From this it is easy to find the *neutral* elements of the two operations and to verify that, with the definitions given, the conditions of distributivity are satisfied. Consequently the polynomials constitute a field. Other examples are the algebraic numbers, the congruences modulo a prime number, Hensel's "p-adic" numbers, and Veronesse's formal series.

It is important to note that the concept of field explicated, defined and designated by Dedekind, had already been used by Abel and Galois. The former had used them in his well known study of the equations that are today called Abelian by showing that they cannot be solved by radicals. Abel defined the notion corresponding to a polynomial irreducible in the field generated by the coefficients of an equation. But the notion of field *as a set* does not appear in Abel's work, or in Galois's.[11] Both of them consider the elements of the set and define them with precision, but the set itself does not appear explicitly in their work.

We encounter here a historical situation similar to the one analyzed at the end of section II, with respect to Gauss, Ruffini, and Cauchy. In addition, as we have seen with the discovery of group structure, Galois introduced the transoperational stage as far as the evolution of solutions for algebraic equations is concerned, a step which Gauss and his contemporaries had been unable to take.

Nevertheless, in respect to the development of the notion of field, Galois remained at the interoperational stage. Dedekind took the next step when he succeeded in identifying and thematizing the structure of algebraic fields, thus ushering in the transoperational stage. In the conclusion to this chapter we shall return to the problem of the relativity of the notions intra-, inter-, and trans-operational.

IV. THE LINEAR INVARIANTS

Felix Klein unified two different approaches: non-Euclidean and projective geometries. For some time to come after this unification, the search for the invariant properties in linear transformations was one of the preferred topics for mathematicians. The search for geometrical properties had become a search for algebraic invariants. Cayley, Sylvester, and Salmon were the key figures during this period, although other mathematicians have done research on the invariants of particular algebraic forms (binary quadratic, ternary cubic, etc.). The discovery that the invariants of certain algebraic forms were, in turn, algebraic forms with invariants led to the formulation of a more general problem: finding a complete system of invariants for a given form.

After a few unsuccessful attempts and certain partial demonstrations, Hilbert resolved the problem. His "basis theorem" shows that for any form (or system of forms) of a given degree and a given number of variables there is a finite number of rational integral invariants and covariants (the basis) such that any other rational integral invariant and covariant can be expressed as a linear combination of the basis. Thus, such a basis defines a complete invariant system of a given form.

Hilbert exhausted in some sense the problem of the search for projective invariants. After having demonstrated the preceding theorem, he further demonstrated that the linear invariants do not completely solve the projective classification of plane curves, since certain types of curves with different particularities cannot be distinguished from each other by their invariants (for example, the irreducible curves of degree 5 having a quadruple point versus those having a triple point with a single tangent).

Now, long before work on the topic of projective invariants had reached a dead end, other results furnished by analysis had opened up new avenues to overcome the problems which the projective methods had left unsolved. Essentially, this concerns the study of new *transformations* leading to new forms of invariants. Let us examine in detail how this evolution took place, choosing as our central topic the theory of algebraic curves and surfaces.

V. BIRATIONAL TRANSFORMATIONS

The development of the theory of algebraic curves began historically with the theory of elliptic functions, that is, the theory of the integrals of rational functions on an elliptic curve. The name "elliptic" was applied to integrals that permitted the calculation of the length of an arc of an ellipsis. Later it was used for all curves ("elliptic curves") yielding elliptic integrals. Then it was noticed that the basic properties of elliptic integrals could be generalized to the integrals of arbitrary algebraic functions. Finally, one began to study these.

The point of departure for this development can be found in Abel's 1926 text, which is at the origin of the theory of algebraic functions. Now, it was not until Riemann that this theory was to have an important repercussion on the geometry of algebraic curves. Riemann's contribution is described with clarity by F. Klein in an address in Vienna, in 1894,[12] at the occasion of the opening of a scientific congress:

> The study of algebraic functions essentially comes down to that of algebraic curves whose properties constitute the object of geometricians, either of the adepts of analytic geometry, where the formulae play the principal role or those of synthetic geometry in the sense of Steiner and Von Standt, where the manner in which the curves are generated constitutes the object of study, using series of points or bundle of rays. Essentially, the new point Riemann introduced into that theory is that of a general (unique) transformation. From this point on, the algebraic curves with their tremendous diversity of forms can be grouped into larger categories. This makes it possible to approach the general study of properties common to all curves grouped together regardless of the special properties of particular forms of the curves.

It should be pointed out, however, that in this aspect of his work Riemann suscribes entirely to the point of view of the theory of functions. In the best known of his papers on this topic ("The Theory of Abelian Functions"), Riemann rarely uses geometric language. Dieudonné describes this as follows:

> It is not one of the smallest paradoxes in the work of this prodigious genius, from which algebraic geometry emerged entirely renewed, that he almost never treats algebraic curves: it is from his theory of algebraic *functions* that was developed the entire birational geometry of the 19th and beginning 20th century.[13]

Riemann's central notion was that of "rational substitution" (later called "birational transformation").

Clebsch was most responsible for the further geometric interpretation of Riemann's theory of algebraic integrals. Clebsch introduced the definition of genus as a basic concept for the classification of curves. For a curve of degree n with d double points, the genus is defined by the expression:

$$p = \tfrac{1}{2}(n - 1)(n - 2) - d$$

In a fundamental study, Clebsch and Gordon (1866) demonstrated that the genus of the curve defined by $f = 0$ corresponds to the number of first-class integrals which are linearly independent. On the basis, they showed that the genus is invariant under birational transformations.

On the basis of this work, birational geometry would burgeon during the last decades of the nineteenth century. New results were obtained by Lüroth, who showed that a curve of genus 0 can be transformed into a straight line by a birational transformation. Clebsch himself demonstrated that the curves of genus 1 can be transformed, in analogous fashion, into curves of the third degree.

Brill and Max Noether completed the work of this school by defining the geometry on an algebraic curve in the projective plane as the set of properties that are invariant over birational transformations.

The birational geometry of that period can be viewed as an attempt at synthesis between projective geometry and Riemann's ideas. Although several of these demonstrations are of an algebraic nature, the methods used are firmly gounded in the theory of functions. The latter is the source of the concepts used by birational geometry. The purely algebraic consideration of invariants would, however, progress very slowly.

The birational tranformations have made it possible to study the properties of algebraic curves and surfaces with a high degree of generality. In the specific case of the study of the singularities of an algebraic curve the following two theorems were derived:

- An algebraic curve can always be transformed, by means of a birational transformation, into a curve free of singularities and, following this, by projection onto a plane curve having only ordinary double points.
- An algebraic plane curve can always be transformed by a Cremona transformation into a plane curve having only ordinary multiple points.

In the case of algebraic surfaces, the reduction theorem—analogous to the first cited here—states: any algebraic surface f can be transformed birationally into a surface in S_5, free of singularities, and thus, by projection, into a surface in S_5 having only ordinary singularities.

As for the second theorem established for algebraic curves, its analog for algebraic surfaces is formulated as follows: any algebraic surface f in S_3 can be transformed by a Cremona transformation, into a surface having only ordinary multiple curves (that is, with distinct plane tangents free of singularities and having a finite number of ordinary cusps); furthermore, any two multiple distinct curves have, in any of their common points P, distinct tangents, and P is never a basis point of polar curves of y.

VI. ALGEBRAIC CURVES: FROM TRANSFORMATIONS TO STRUCTURES AND CATEGORIES

At the end of the nineteenth century the theory of algebraic curves was to reach a new stage, due in large part to Hilbert's work on the rings of polynomials.

The structures of rings and of ideals had been known and applied, as we have already shown, by Kronecker and Dedekind, even though the term "ring" was introduced by Hilbert. Algebraic geometry, which begins with them, makes systematic use of these concepts. Brill and Max Noether, for example, in the thesis cited above, demonstrate the birational invariance of the linear series defined on a curve by means of the theory of rings generated by polynomials.

The clear and precise translation of the geometrical problems into the terms of the theory of ideals became possible only after Hilbert had proved his famous theorem of the finite basis and the theorem known under the name of Nullstellensatz (theorem of zeros). For example, the problem of defining an irreducible algebraic left curve as the intersection of a finite number of algebraic surfaces has its analog in the algebraic varieties with which one can always associate the ideals of the rings of polynomials: the set of zeroes of an ideal (that is the set of points at which all polynomials of an ideal cancel out) is the intersection of a finite number of algebraic hyperplanes.

The result comes from Hilbert's theorem according to which such ideals accept a finite system of generators. Max Noether and E. Netto had obtained partial results with the inverse problem: given a set of polynomials $F_1, F_2, \dots F_n$, the problem is under what conditions a polynomial cancels at the points of the variety defined by the equations $F_1 = 0, F_2 = 0, \dots F_n = 0$. Hilbert's zero theorem showed that for any polynomial F of this variety, there exists a whole number h such that F^h belongs to the ideal A defined by the given polynomials. Now, the problem of the decomposition of ideals was still unsolved. Lasker took a decisive step by formulating the conditions under which a polynomial can belong to the ideal generated by the n polynomials $F_1, \dots F_n$. He did this by introducing the notion of primary ideal and showing that in the formulation given above, the set of polynomials that satisfies the condition F^hEA is the intersection of a finite number of primary ideals.

During the second decade of the twentieth century, the theory was completed by Emmy Noether, who succeeded in rigorously redefining the problems within the framework of abstract algebra. E. Noether's theses completed a stage in the development of algebra, by unifying the theory of integral algebraic function (polynomials) and the theory of the ideals of whole algebraic numbers.

The following stage began just before 1920, with the notion of local rings—(those which possess a single maximal ideal). That structure was first studied from a purely algebraic viewpoint. It was not until after about a decade of research on these algebraic properties that the structure was clearly applied to the theory of algebraic curves, thanks to the work of O. Zariski. Zariski used the method of birational transformations, although in this case the elements of the birational transformations were more complex structures than those that had been used during the apogee of birational geometry during the last century.

Zariski used a method of "localization" based on the introduction of an appropriate topology (today called Zariski topology), which permits the attribution of a "local ring" to each point of an algebraic set. From this was obtained a method which permits one to associate with all varieties a "normal" variety (that is, a variety such that the local ring of each point is integrated and integrally closed). Zariski thus came up with a method of solving singularities through a process of normalization associated with a birational transformation.

The important step on the way toward complete algebraization of algebraic geometry was the transition from affine (or projective) to abstract varieties. The former had been defined by systems of equations—that is, by polynomials; in this way, a correspondence is established between "geometric objects" so defined and a certain type of ring. Properties of the affine variety were thus reflected in the ring which was associated with it in invariant fashion.

With the introduction of the abstract varieties one is no longer limited to the arbitrary choice of a system of equations, since the same variety can correspond to different systems of equations. At the same time, the necessity to immerse these "geometrical objects" in a space (either affine or projective) disappears.

The development of the concept of abstract variety would start with an arbitrary commutative ring with an identity. A "geometrical object," which now is simply a certain class of elements endowed with a topological structure, would be joined to it. The class in question is that of all prime ideals of A, which is called the "spectrum of A" and represented by Spec. A. The prime ideals are called points of the spectrum. When A is the ring of subsets of an affine space, Spec. *A* has a clear geometrical interpretation in terms of points, irreducible curves, and surfaces. Spec. A thus possesses a special topology, called spectral topology, so that it is possible to refer to Spec. *A* as a topological space. The geometrical object associated with a commutative ring *A* is thus identified as a set of elements with topological structure.

The definition of a structural bundle on topological space Spec. *A* and the application of the notion of ring space (topological space on which is given a bundle of rings) lead to the concept of schema. This concept defines a category which is equivalent to the dual of the category of all commutative rings.

The concept of schema introduced by Grothendick, in his reconstruction of algebraic geometry provides the means for classifying the "geometrical objects" in a very general fashion. As Dieudonné pointed out, "this led to the kind of synthesis that Kronecker had been the first to dream of," in as much as the schemata can be applied to the algebraic varieties as well as to the theory of numbers. At this point a level is reached where transformations, as a basic concept, which characterized the long period of birational geometry, was replaced by abstract structures. The concept of morphism was thus to play a dominant role so that the properties of schemata finally came to be replaced by those of morphisms.

VII. CONCLUSIONS

We indicated in the introduction to this chapter our intention to trace the development of algebra by concentrating on the sequence of intra-, inter-, and trans operational stages. In geometry, we had no difficulty in differentiating these stages. Euclid, Poncelet or Chasles, and Felix Klein were the key representatives of each of these stages. In algebra we note the following anomaly: once constituted, its domain remained limited for centuries to a fairly restricted topic—algebraic equations—which later on turned out to be only a restricted partial aspect of its proper domain.

In spite of the analogy just noted, it is possible to uncover a strange inverse relation in the development of geometry and algebra. Geometry began with a well defined identity and a precisely delimited domain, but in the course of its development it gradually lost this initial identity to the point where it became algebra. In contrast, algebra began with a partial and very restricted definition and gradually acquired its true identity up to the moment when it surfaced as the study of structures, several centuries after its inception.

We have carried out a preliminary exploration of the origins of algebra in order to determine how it established itself as a branch of mathematics. The conclusions we reached—supported to a large extent by Jacob Klein's studies—can be summed up as follows:

a. Viète carried out the transition (which the Ancients had not been able to realize) from the concept of "arithmos" to that of general symbols, upon which algebra came to be constructed as a new discipline.

b. To achieve this, Viète performed a synthesis between the geometrical analyses of Pappus and the arithmetical methods used by Diophantus.

c. Even though the concepts of transformation and invariant were not made explicit (thematized) during that period, they undeniably play a fundamental role, since they made possible the transition from the concept of symbol, as used in Antiquity, to represent in a general way, a concrete number to that of a general symbol as a form representing a general number (that is, any number whatever).

When algebra first established itself as an independent discipline, its central theme was the solution of equations. In this case, we were able to differentiate the three levels characteristic of its development.

During an initial, extremely long, period the only concern was the search for solutions to specific equations. The methods used were purely empirical, trial-and-error. Each equation was treated as a separate object. This is undoubtedly an intraoperational period.

It was only in the eighteenth century that mathematicians began to look for more general methods and, more importantly, to formulate the general problem concerning the existence or nonexistence of solutions. The transformations of equations that allow the reduction of an unsolvable form to one that is solvable largely dominated mathematical research. As in the case of geometry, analysis was to play a decisive role in this process. Lagrange and Gauss were the important figures of this period, which from our point of view, constitutes an interoperational period.

With Galois and the development of the theory of groups—the first thematized structure in mathematics—the period of equations and their solutions came to an end, and a new one began, during which structures predominated. This was a long transoperational period.

From this point on, the process becomes considerably more complex. We have limited ourselves to a few developmental trends that seemed sufficiently representative of the general characteristics of the total process. Each of these trends reveals the same mechanisms that were identified for general development: interrelating internal properties followed by comprehension of these properties as invariants in transformations; discovery of transformations followed by the interpretation of such transformations as manifestations of a total structure from which they can be derived as intrinsic variations. All this is accompanied by a relativization of concepts and reinterpretation of variables. These are the mechanisms which determine the stages characterised as intra-, inter-,. and transoperational.

The role played by these notions in the process of conceptualization and construction of theories encompasses three different but related aspects, which need to be made explicit before we can give an epistemological interpretation of these notions. These aspects concern: (1) the successive levels [stages] in historical or psychological development; (2) the different phases within each of these levels, since these require a regular sequence of sublevels for each new construction; (3) the way in which previous acquisitions are reinterpreted from the perspective of the newly attained stage.

We have already referred to point 3 in chapter 3, on the development of geometry, by showing how at each stage there is a reorganization of the knowledge acquired during the preceding stage. As for points 1 and 2, we should add that when two subsystems are coordinated so that a part common to both or an intersection is discovered, the level attained by this common part may be superior, equal, or inferior to that of the subsystems constituting it. This fact supplements the general rule according to which each stage repeats the total process in its sublevels—that is, as we saw in chapter 3, a sequence of sublevels of the type intra, inter, and trans. That is, one finds the same process at work in the phases of construction toward a higher level as in the sequence of the main stages. While keeping this in mind, let us review the evolution of the group structure, which is in itself transoperational in character:

a. The first groups introduced by Galois concerned only permutations already given and thus not constructed by the group itself: from the group's point of view, the relations involved were thus intraoperational, even though from the point of view of the entire stage this is transoperational.

b. After that came, with Klein, the groups of transformations which play a constitutive role, such as the projective transformations which represent in themselves components of the group. This structure is thus of the interoperational variety.

c. The concept of abstract group was elaborated referring to any class whatever such as that operating on vector spaces. This finally represents the transoperational with respect to the specific elaborations concerning the structure of the group in general.

We already indicated that underlying these apparently purely descriptive concepts intra, inter, and trans, we expect to be able to uncover some constructive mechanism.

But first it will be useful to give a more detailed description of the progressive constructions and their retroactions. Let us assume the following three stages *A, B, C*. Let us suppose that *B* has the sublevels *a, b,* and *c*. Now even at sublevel *a* of *B*, that is, at an intralevel, there is already reinterpretation of *A* in the sense that *A* becomes subject to the type *B* operations, while elements of *a* of *B* will be involved in the transformations of the "inter" type at sublevel *b*, and so on. This shows how the sequence intra, inter, and trans can be found, in a proactive sense, both in the sequence of sublevels and of stages, so that it has a retroactive effect on preceding constructions due to reorganization. This becomes possible as a consequence of the new constructions. On the other hand, this kind of reorganization is also a necessary condition for the constructive generalization of previous constructions. In fact, "constructive generalization" does not consist in assimilating new contents to already constituted forms, but in generating new forms and new contents—i.e., new structural organizations.[14]

The notion of a sequence of three stages, of the intra, inter, and trans variety, poses epistemological problems which necessitate a detailed analysis. One of the problems that arises is that of the late appearance of transformations and structures. In referring to the notions elaborated a long time ago by genetic epistemology, we can account for the early emergence of elements and the late appearance of structures in terms of progressive mechanisms of abstractions and generalizations: structures require both a greater degree of reflective (or thematizing) abstraction and a more nearly complete generalization.[15] However, this interpretation remains inadequate; it does not account for the principles motivating the evolution of such constructions because in the final analysis, the abstractions and generalizations are still more than instruments.

As for the successive and ultimate finalities of the process leading from the intra to the inter and then to the trans levels, the answer seems to be fairly obvious: neither the mathematician nor the child at a certain level are ever satisfied with simply observing and discovering (which means inventing things): at each stage, they strive to attain the "reasons" for what they find. This then comes down to a search for the "necessity" inherent in the generalities and relations established, since the subject does not accept, in the final analysis, a construction as being valid unless it acquires intrinsic necessity through explicit reasons.

It now becomes possible to explain the general process, which generates the various forms of necessity as they develop and whose acquisition proceeds only gradually. The relations between elements proper to the intra level either have no necessity or, if they do, they achieve only very limited forms that are still very close to simple generality. Understanding states as resulting from transformations and, to this end, performing local transformations of elements—as is the case at the inter level—provides a first access to the necessary connections which intrinsically determine their own reasons. But transformations, in turn, require an explanatory rationale, and the search characteristic of the trans level—which leads to structures—is the response to this new need, since a total system of transformations generates new transformations and provides the reasons for their systemic composition. But it is clear that this "total" character itself remains relative and that the movement continues, as we have seen, among other things, with the transition from structures to categories of structures. In a word, the sequence from the intra to the inter to the trans levels is merely an expression of an identical process. On the subjective side it is the search for explanation; on the objective side it is the achievement of necessity, which always remains relative, but which increases constantly from one stage to the next.

VI

The Formation of Pre-Algebraic Systems

Algebra is the science of the general structures common to all branches of mathematics, including mathematical logic. But to get to these structures, two preliminary stages were necessary, as we recalled in our historical chapter. The first can be characterized as "intraoperational," since only particular systems were analyzed, considered in terms of static and limited properties, such as the theory of geometrical proportions elaborated within the Euclidean perspectives. The second achieves an "interoperational" level with Viète's analyses directed at transformations, which were made possible by an abstract and general symbolism. Quite a long time after this are constituted the "transoperational" syntheses, which led to "structures" in the contemporary sense of the term. The first constructions of this kind are found in Galois's work. One of the characteristics proper to all three of these successive systems of theoretical elaborations is that they all use, as psychological instruments, what may be termed "reflective thematization"—that is, an exhaustive conceptualization of progressively constructed mathematical objects, and this even before such representational intuitions were developed into axioms.

Now, since these "objects" are all products of the subject's activity and not given as physical experience, the problem naturally arises as to where to seek the source of these activities. Endogenous conceptualizations cannot be creations *ex nihilo*. They require as a prerequisite a certain "knowhow" in action that is not thematized, not even conscious in the details of its articulation. In particular, this means that in the "natural" thought of children and adults who are not mathematicians, actions and manipulation of objects are not built up by chance, but are organized, sometimes quite systematically, in the form of what one might call "pre-structures" or pre-algebraic systems. The simplest and the best known examples of such systems are such groupings as classification and seriation. Now, these elementary systems are of epistemological interest not so much in that they prepare, on a pre-scientific level, such advanced forms of organization as the "group" and the "lattice" (although this is partly true), but in that they present three stages analogous to the intra, inter, and trans levels found in the evolution of logico-mathematical thinking.

The main problem is then to determine whether we are dealing here with the same process in spite of considerable differences in levels. The three periods distinguished in history involve, in fact, centuries of evolution, from antiquity to Viète and from Viète to Galois, while in the case of similar sequences in psychological development, we are dealing with extremely rapid developments (from four or five years of age through eleven or twelve). Also the degree of differentiation between the various levels is much less complex. Is it then legitimate to consider our triad intra, inter, and trans as a common mechanism that, in the manner of reflective abstraction and completive generalization,[1] is necessarily operative in all domains and at all levels? Or are our proposed comparisons merely convenient ways for the historian and the psychologist to describe quite disparate facts in a language that facilitates analysis?

The problem did not arise in respect to the evolution of geometries, as it might have, because the contrast between the intrafigural (where only properties internal to a figure were considered) and the interfigural (the surrounding space) was so obvious. Therefore, the comparisons between history and psychological development did not appear so artificial. In the domain of algebra, however, the contrast between theory and practice appears wide indeed. The observer and researcher of psychological development, whose laws are unknown to the subject, will note the division between theories (developed in the course of history) and the practical organization (largely unconscious) involved in the subject's formation of structures and prestructures. This may, at first glance, make our triad seem unduly artificial.

Now, this sort of question calls for two kinds of answers. We shall examine the principal one below. If this triad really constitutes a common mechanism of a very general nature, we shall show, it is as a result of internal necessities, and not simply a regular sequence. In addition, each stage encompasses sublevels, which follow the same sequence and for the same reasons. This fact is of fundamental importance. We can give as an example the third of the great historical periods, which we characterize in terms of the elaboration of very general "structures," designating them collectively as transoperational. Now, it is easy to see, as we have noted in the preceding chapter, that this trans stage is in turn subdivided into three sublevels. One might call them "trans-intra," "trans-inter," and "trans-trans." The last of these three results from the application of new operations to earlier structures; now, these newly thematized operations, which are derived in a sense from previous ones, consist in "abstract groups" and lead to structures of an even higher rank.

If our triads thus contain nested-triadic sublevels, then there is no reason why this should not also be the case during that important prealgebraic period when the subject, as yet incapable of systematic thematization, nevertheless constructs, on the level of action and practical knowhow, what the observer can interpret only as a progressive formation of structures. Therefore, we shall attempt to identify our triads in the domain of psychogenesis, using the same terminology, but with the proviso that we are dealing here only with actions. The terminology used, then, serves only to describe the phenomena in our language; it does not prejudge what the subject in fact consciously knows about it.

I. INTRA, INTER, AND TRANS AT THE LEVEL OF ACTIONS

Our extensive research on the development of operations in the child has led us to distinguish three successive periods in all the domains explored so far:

1. *Preoperational.* In the course of this stage repeatable actions are gradually built up. These modify the objects but neither undergo transformations nor are coordinated with each other.

2. *Concrete operational.* In this stage, operations become organized into systems ("groupings") which include certain transformations of the operations themselves.

3. *Hypothetical deductive operations.* Here, there is a synthesis of transformations, which can sometimes even take the form of "groups."

Thus, it is clear that the three stages representing age levels four-to-five, seven-to-ten, and eleven-to-twelve (and beyond), respectively, correspond to our sequence intra, inter, and trans. We shall show this by means of concrete examples before going into the reasons which render such a progression necessary and which explain why there are three stages (in the manner of the "thesis," antithesis" and synthesis of classical dialectics) rather than some other number.

A. The intra is characterized by the discovery of some operational action and the analysis of its internal properties or immediate consequences. There are two limitations at this level: an absence of coordination of this preoperation with others into an organized "grouping," and a presence of errors in the internal analysis of the operation (which have to be successively corrected) as well as gaps in the consequences subjects are able to deduce.

To demonstrate this "intra" situation, a good example is the experiment in which the child is presented with two identical beakers, one *(A)* transparent and the other *(B)* hidden by a screen. Both contain a few pearls identical in number so that $A = B$. The subject is asked to add pearls one by one to each beaker. The question is then whether, after a few such additions, $A + n = B + n$. All subjects, from four to five years on, are convinced, as long as n is small, that the equality $A + n = B + n$ is conserved for the pearls already in the beaker. This is the basic operation, which is thus performed successfully. However, it remains to see if it is coordinated with other operations or, more simply, if it can be applied to accurately deduce the consequences implied by it. Now, even one of the simpler kinds of consequences leads to very revealing fluctuations: when we ask simply what would be the outcome if one continued to proceed in this way for a certain time (say "an hour" or "till the evening"), most subjects refuse to make a decision ("one cannot know without counting," etc). There are a few rare advanced cases, like that of the five-and-a-half-year old who answered, "if you know it once you know it forever." This remark appeals to the notion of recurrence characteristic of the inter level.

In another problem, n pearls were placed in A and m ($>n$) in B, where m is slightly greater than n (by about 2 to 3 elements). We ask if this inequality remains the same if one adds successively one pearl to each of the two beakers, or if in that case the two collections will eventually be equal. Here again the answers vary and subjects tend to believe that the inequality will eventually disappear because the quantities added to each side are constantly equal. Let us add that if the two collections A and B are very unequal and n elements are removed in pairwise fashion, subjects frequently believe that the n elements taken from the larger collection "make more" than those taken from the other, because of a confusion between extension (number) and intension (large or small collection).

These facts may suffice to characterize the intra stage, which consists, then, in a concentration on a repeatable action or a correct operation, but without the capacity to insert it in a system of conditions or consequences, which would extend its application and include it in a system of interdependent transformations. In fact, there is already the beginning of transformations, but applied only to the objects modified by the actions involved, rather than to the initial actions or operations. These remain isolated and analyzed and comprehended only in terms of their properties taken individually and irrespective of others; hence the analogy with the 'intrafigural' of chapter 2, where a figure is characterized in itself without reference to the surrounding space.

B. The "interoperational" stage on the other hand, is characterized by the capacity to deduce from an initial operation, once it is understood, others which are implied by it or to coordinate it with other similar ones. This leads to the constitution of systems including a certain transformation, which is new; but there are still fairly limiting restrictions on the compositions possible—

i.e., the limitation to a strictly step-by-step approach. Thus, in the domain of classifications, subjects are able to combine within a class A all the objects presenting a characteristic a, and also to include this class A within another class B, larger than A, which contains all the A and also the elements B that are not A, and which we designate A' (hence $B = A + A'$). They can even go further and include B within C following the same principle ($C = B + B'$), etc. But what cannot be done at this level is to go beyond such "natural" embeddings and to combine, directly in a single class, elements that are far removed from each other, such as a fly and a camel, without going through a complex set of intermediate embeddings, or else to stay with very general classes, such as animal or vegetable; in the latter case, the subclass structure fails to be observed. These kinds of limitations force us to distinguish the interoperational systems from the transoperational "structures," that is, the step-by-step coordinations from true syntheses. It is in order to mark such differences that we introduced the notion of "groupings."

Nevertheless, within groupings certain kinds of distinct compositions are already possible. Thus, within the groupings of genealogies (the pedigree without taking account of marriages) the central relationship of son to father generates those of brother (son to the same father), grandfather, cousin (grandson to the same grandfather but not son to the same father), etc. Hence, a set of equivalence relations such as the one that relates this characterization of cousin to another one that describes it in terms of father's brother's son (uncle) or father's nephew, etc. But these are still only step-by-step coordinations, hence related to discursive[2] processes. They do not yet attain those "transcending" capacities that are characteristic of the trans level. These are the result of syntheses proper.

However, it should be stressed that, at this intermediate level of groupings, where relations of the "inter"-variety are being formed, the latter begin to develop further and further in the direction of transformations. Thus, comparing the set of groupings relative to classes to those of relations, we even find two general systems of transformations. However, these are not yet combined in a synthesis, as will be the case at the trans stage. One of these general transformations—so apparently simple that one might almost doubt that it is a transformation—is negation, which is particularly important in the groupings of classes. We shall discuss below certain difficulties related to it. The second is constituted by reciprocities. It concerns particularly the groupings of relations and also causes certain difficulties. Let us then examine these two kinds of transformations. It is immediately clear that they constitute the two possible forms of reversibility—that is, the common characteristic of all operations and their compositions, hence that which essentially marks the progress between intra- and interoperational systems.

As for negation, it constitutes the condition *sine qua non* for the delimitation of all embedded classes, since, if $A \subset B$ this means that (if $B > A$) $B = A - A'$ and that $A = B - A'$. In other words, *omnis determinatio est negatio* (all determination is negation), as Spinoza said. This is true in both directions, since $A' = B - A$, so that in every case, the implication $(p \rightarrow \bar{q}) \rightarrow (q \rightarrow \bar{p})$. All this may seem perfectly obvious and even in some way accessible to direct observation. Still, our earlier research on the quantification of class inclusion[3] clearly shows that these transformations only come to be regulated gradually during the inter stage, while they are generalized not at all during the intra stage. For instance, we presented children with ten flowers of which six or seven were primulas, the others any other kind of flower. Subjects agree readily that all these are flowers and that the primulas also are flowers. But when we ask them if in a bunch of flowers, "there are more flowers or more primulas," the usual answer, up until the inter level is that there are more primulas because there are only three or four daisies, etc. In spite of all our efforts, subjects keep on reasoning as if the total set of ten flowers was split into two parts: the subclass A no longer belongs to the superclass B in the form of $A = B - A'$, while the superclass B is reduced to what remains, i.e. A'. To be sure, the opposition A versus A' implies an elementary negation, since the A are not A'. But this negation is not yet relative to its referent B, since the latter ceases to exist following the partition. Now, the negations involved in classification are all relative to a referent ($A = B - A'$; $B = C - B'$; etc) and for this reason they constitute one of the conditions necessary for embedding. The absence of embedding of *A* within *B*, which prevents the quantification of inclusions as in the experiment just cited, is therefore related, in one of its essential aspects, to difficulties in the regulation of negations.

As for reciprocities, the way in which they constitute general transformations is more immediately evident in the domain in which they predominate—i.e., that of the groupings of relations. Once again it can be noted that their acquisition does not occur all at once during the inter level. Rather they require progressive construction. At the intra level, the absence of reciprocity goes as far as total negation in the case of the elementary relation of "brother":

"Do you have a brother?"

"*Yes, F.*"

"And F, does he have a brother?"

"*No, we are only two boys in the family.*"

Similar difficulties appear later with the relations "cousin," "nephew" versus "uncle," etc. More complex reciprocities appear, of course, between interdependent but antagonistic variables, as in the case of action and reaction in mechanics.

C. The transoperational level is easy to define as a function of what precedes. There are not only transformations, but also syntheses between them, leading all the way to the building of "structures," albeit only at the level of actions and without thematization. The most noteworthy of these structures unites inversions and reciprocities—i.e., the two forms of reversibility. It includes the total set of parts, not limiting itself to disjoint relations or classes to be structured. The "structure" attained in this case is an authentic group, which we call INRC, since for any operation like $p \rightarrow q$, we can leave it identical (I), invert it to negation $(N) = p \cdot \bar{q}$, transform it to its reciprocal $R = q \rightarrow p$ $(\bar{p} \rightarrow \bar{q})$ or to its correlative $C = \bar{p} \cdot q$, hence $NR = C$, $NC = R$, $CR = N$, and $NCR = I$. Obviously, this formulation is only that of the observer, but subjects from the ages of eleven or twelve on construct such syntheses by composition in actions and by progressive inferences in situations where it is necessary to coordinate negations and reciprocities within a unified system. A simple example is that of the balance. One can increase or diminish the weights while leaving them in the same place, or compensate a weight located near the center by one of less weight located at a greater distance. Another example is that of relative movements where it is necessary to coordinate the two referents, etc.

But we still have to examine two more problems. The first is only terminological. We decided to limit the term "structure" to those which include a synthesis of distinct transformations, such as the inversions and reciprocities just discussed. Here we are obviously dealing with "transoperational" systems, because the limits of each of the constitutive transformations are transcended. But we also considered the "groupings" as already constituting structures, and this remains valid even though they remain rather limited by virtue of their purely step-by-step compositions. (Hence the notion of "immediate successor" used by Wermus in his axiomatization of groupings.) It is thus sufficient, in order to avoid all ambiguity, to distinguish structures by generalization and iteration, which are already elaborated at the inter level, from the synthetic structures of the trans level; here we use the short form "structure" to designate the latter term.

But the essential problem that remains to be solved is that of the motor which drives this change from the intra to the inter and finally to the transoperational levels. In addition, there is the problem of the relations between the constructional dynamics and the general process, leading from the simple use of an operation to the construction of "operations on operations." To facilitate the discussion it may be useful to recall certain facts which show what the subject is able to make use of in the simplest groupings (seriations and classifications) at intermediate levels of functioning between the grouping characteristic of the interoperational level and the syntheses of the transoperational level.

II. ON SERIATION AND CLASSIFICATION

If one presents subjects with a number of sticks of unequal length and asks them to "arrange" the sticks, even preoperational subjects at the intra level are able to construct a series $A < B < C$ . . . as a kind of "staircase," but they cannot imagine analogous series with other objects—for example, circles of different diameters. Thus, they remain at the intraoperational stage, where a form remains local and restricted to certain contents. In contrast, interoperational generalization occurs when seriation is used as a cognitive instrument to analyze various contents such as a series of polygons with an increasing number of sides, etc. As for the "transoperational" level, it consists in combining different seriations with the aim of deriving from them a richer and more "powerful" operation, as was the case in the following well known study. Subjects were presented with small fish of various sizes and a certain number of food balls also of various sizes and were told to distribute these as necessary to meet the food requirements of the fish. In this situation, subjects who had reached a certain level seriated the fish according to their sizes and the food balls similarly, then established one-to-one correspondences between the two series, which to them represented the simplest form in which they came to discover the notion of proportion.

These facts, simple as they may be, show that an operation, once constituted and no matter how close it remains to everyday activities like arranging a sequence, does not remain inert and isolated (intra) very long, but sooner or later comes to be structured in the direction inter and trans; these, then, become infinitely extended until true structures are formed.

Another example is provided by classification. The most obvious form of intraoperational relations consists here in grouping similar objects in a class $A'1$ (hence the possibility of building the series $C = B + B'$, *etc*). Progress in the interoperational direction comes about when the subjects discover that the grouping of B into $A1$ and $A'1$ is not necessarily the only one possible and that the same elements can be distributed using different criteria. In this case, the same class B will be composed of subclasses $A2 + A'2$ such that $A2$ belongs to the earlier set $A'1$. Hence, the elements $A'2$, complement to $A2$ within the class B, comparable to $A'1$, complement to $A1$ (= the "others" with respect to the initial set $A2$ or $A1$), are no longer the same "others" so that this term acquires a relative sense. We call these changes in classification "vicariance" with respect to an identical set of elements. The grouping in vicariances is important particularly in the elaboration of kinship relations where, as we saw, the interoperational has a definite advantage over the intraoperational.

Vicariance is of interest chiefly in the following way. While it is true that vicariances come to complete rather well the first classificatory behaviors, initially they consist in only successive modifications of an initial classification, and no overall simultaneous combinations of different possibilities are evident. For example, a number of stimuli can be classified according to shape, size, or color, but subjects do not see the possible intersections between the three kinds of subclasses, which they construct only in succession. In contrast, they later experience little difficulty in finding as many "small red squares" as "big blue circles," etc. or in realizing that one of the subcollections is larger than the other. Now, this capacity for differentiation and simultaneous embedding leads sooner or later to the 2^n combinatorial—that is, to the "total set of parts," but in action only, without mental computation or thematization.

The two examples we have just cited have a common characteristic, which is of fundamental importance in the construction of pre-algebraic action systems: this is the process leading from simple relations to relations between relations. Given two series $A1 < A2 < A3 \ldots$, and $B1 < B2 < B3 \ldots$, both of which are based on a relation of size, $x < y$, if these are related to each other to yield quantitative equivalences of the form $A1/B1 = A2/B2$, the composition of these two kinds of relations leads to the system of proportions, which is of essential importance and which gets generalized at about eleven or twelve years. In the case of the combinatorial, which is the set of all parts (attained at about the same age), the composition of the relations involved even takes a form that can be found in many instances: an operation raised to some power; in fact, a combinatorial is essentially nothing but a classification of all classifications possible for a given array of elements.

Let us note in this respect that this is the way numerical multiplication gets constituted, which is understood a good deal later than addition, as it results from an addition of additions and involves two phases. During the first phase, the expression $m \times n$ (for example 3×4) is understood simply as an addition of some elements m and only the result is noted. In contrast, during the second phase, subjects become conscious of the number of times *m* was added distinguishing thus the multiplier from the multiplicand ("I've taken *m* times *n*"). This promotes multiplication to the rank of a new operation to yield a synthesis of additions.

III. THE NATURE OF INTRA, INTER, AND TRANS

It is obvious that this triad constitutes a dialectical sequence, but this does not explain anything as long as the characteristic process involved, by which the instruments of change become superseded in turn, remains unspecified.

Let us first adopt the following notational convention for the sake of facilitating the presentation: *Ia* is to stand for intra, *Ir* for inter, and *T* for trans. The triad is thus represented as *IaIrT*. This being established, the first thing to note is that this triad always involves the construction of operations on operations, but that the inverse is not necessarily the case. Let us also note that the sequence *Ia, Ir,* and *T* must necessarily follow this order, since the elaboration of *T* as a system of all transformations combined to a totality with new properties presupposes the formation of certain of these transformations at *Ir,* which in turn implies knowledge of properties analyzed at *Ia.*

This necessary order of succession implies two kinds of dialectical movements. One of these adds to the properties analyzed at *Ia* the transformations elaborated at *Ir;* the other synthesises in *T* a system of these transformations constituting totals with systemic properties that are novel with respect to *Ir*. But no matter to which contents these successive constructions are applied, constructions which we designate as *Ia, Ir* and *T* include developmental changes from *Ia* to *Ir* and *T* respectively. Such elaborations include, at each level, general psychogenetic or historical processes, which one might summarize as follows. In the first place, a preliminary and necessary phase is the one where particular cases are analyzed—cases not yet related to each other (phase *Ia*). These are then compared to each other and differences and correspondences are found between them which lead to the construction of transformations (*Ir*). Once these transformations are mastered and generalized, new syntheses become possible (*T*). That is, totalities that were inaccessible until then, possessing new systemic properties, are now possible. But let us emphasize that these three phases and their characteristics are functional rather than structural, hence common to all levels of development and not specific to any particular one; in other words, they are inherent in any construction rather than tied to particular domains or developmental levels. Put more precisely, they only describe the psychodynamic aspects of developmental changes in general; they are not directed at any one of these changes in particular.

But if this is so, then it is obvious that our descriptions in terms of *Ia, Ir* and *T* remain necessarily relative. As we pointed out in the beginning of this chapter, each of these main stages includes sublevels so that it is possible to distinguish, within stage *T*, for instance, phases which we have been calling "trans-intra," "trans-inter," and "trans-trans." One clear historical example is that of the theory of "categories"; this theory evidently belongs to stage *T* with respect to the earlier stages of algebra. But, considering the evolution of the theory itself, it is evident that it began with a "trans-intra" phase, where it was limited to the analysis of certain correspondences. Then came a "trans-inter" phase, when transformations between morphisms were studied. Finally, with the discovery of functions, a general theory of categories evolved, which constitutes the "trans-trans" level for this theory.

These cognitive hierarchies include two kinds of structures: proactive embeddings, where domains become extended in the course of successive periods in the construction of knowledge, and retroactive embeddings, where what is acquired at some level n enriches retroactively the relations already established at earlier levels $n - 1$. An example of this can be found in chemistry, where the explanations in terms of electrons have brought about a new conception of valency.

If this is so, one can apparently conclude that the series *IaIrT* does not consist in simple linear changes of the sort that can be found in any elementary dialectic sequence. One has instead to speak of a continual evolution of the very instruments of change. This gives to cognitive instruments their particular richness and complexity. From this point of view, algebra, as theory of forms or structures, constitutes a particularly relevant model for an adequate general theory of intelligence or of knowledge, because only this algebraic interpretation of human reasoning liberates us from the dilemma of empiricism and apriorism. One might even go so far as to claim that the sequences *Ia, Ir,* and *T* have their roots in biology (cf. embryogenesis, etc.): in this way, they come to realize the dream of an integrated theory of constructivism between the initial biological structures and the final logico-mathematical creations.

VII

The Development of Mechanics

In chapter 1, we indicated some of the more salient features characterizing the transition from medieval Aristotelian mechanics to the type that developed in the seventeenth century. Recent historical research (especially the research over the last 40 years) indicates that the rush of scientific activity that developed during the late Middle ages makes it necessary to situate this transition within a perspective rather different from the traditional one—that is, the one which dominated historical research during the nineteenth century and the first half of the twentieth century. That perspective can still be found in certain physics texts.

It is now known that the "scientific method" is not an invention of the seventeenth century. At that time, it already had a long history and had reached a remarkable degree of development. It is true that the scientific revolution contributed largely to the significant progress found in the elaboration of scientific methodology. However, the most important contribution of the period was not in the area of methodology. Nor was the scientific method the source of the evolution of that period. It was subordinated to the world view and the nature of the problems raised. Thus, it is in their world view and the nature of the problems they investigated that we find the fundamental difference between Oresme and Galileo. Crombie expresses this very aptly: "The procedures of science are methods designed to answer questions about phenomena; the questions define the phenomena *and constitute them into problems*" (our italics).

The distinction we drew between an appropriate scientific methodology and the epistemic frame of reference (that is, the epistemic presuppositions which direct the application of a methodology) made it possible to center our interpretation of the transitions that interest us on the changes occurring in the epistemic frame of reference. In this way, the preceding chapters have revealed a remarkable parallelism between the historical and the psychogenetic process in the development of explanations of empirical phenomena. In both psychogenesis and in the evolution of pre-Galilean mechanics, there is a parallelism in the content of thinking and in the mechanisms involved in the elaboration of concepts.

I. NEWTONIAN MECHANICS

From Newton on we are faced with a different problem. The high level of abstraction and complexity characteristic of Newtonian mechanics and its further developments (the mechanics of Lagrange and Hamilton up to the mechanics of quanta) are obviously far removed from the contents one can study in the development of physical thinking in the child and the adolescent. Hence, we shall have to focus, as we did in the case of mathematics, on the mechanisms controlling the process of stage transitions in historical development and on the relations between this process and the developmental mechanisms mediating the transition from one stage to the next in individual cognitive development. For this reason, we shall not refer here to the contents of the concepts used at each of the stages in the evolution of mechanics, except in cases where such reference is indispensable for showing the essential aspects of the structure of a theory and for the interpretation of the significance of a change from one stage to the next.

We shall focus our analysis on Newton's mechanics. The first edition of *Principia Mathematica* appeared in 1686. Newton stands at the apex of a development of which Galileo, Descartes, and Huygens were the chief architects (and without which it would be difficult to understand Newton). But, historians of science are far from being unanimous about the role each of these protagonists played in the scientific revolution.

During the nineteenth and early twentieth centuries, historians considered Galileo as the principal agent of the scientific revolution. According to this view, Newton's mechanics was only a corollary of Galileo's work. However, historical research on the development of medieval science has brought about a certain shift in perspective. Without detracting from Galileo's undeniable genius, it may be said—and this is our point of view—that his work in no way constitutes a spectacular creation without precedent. Rather it is a quantum jump in the development of a tradition whose roots lie in the "quattrocento" in Paris and Oxford, and lead up to Galileo by way of the university of Padua. According to this view, Galileo is a "corollary" of the science of the preceding century, whereas the great innovator and creator of mechanics—called "classical" mechanics today—was Newton.

The demystification of Galileo's work is of considerable importance to our historical approach, that of an epistemological laboratory. To say that Galileo did not perform the "experiment" of letting objects of various weights fall from the top of the tower of Pisa in order to see if they would all hit the ground at the same time is more than a correction of the historical account of this episode, which is a pure invention of Galileo's admirers. To note that Stevin had in fact realized a similar experiment half a century before Galileo, but which had no impact among his contemporaries, is more than rendering homage to a sixteenth-century figure who is still unknown. In each case, the information is very significant for the conception of the history of science as considered from the point of view of an epistemological laboratory. Galileo did not see the need to do the experiment: his conception of the fall of bodies allowed him to infer the results. In contrast, Stevin's experiment—which in itself destroyed to a large extent the medieval Aristotelian speculations—had no repercussions because no conceptual frame of reference existed within which the experiment could be inserted in an explanatory system.

We can now justify more specifically why we decided to focus our analysis on Newton's work: it is because Newton's system came to replace the Aristotelian system in the description and interpretation of the laws of motion. In addition, Newtonian mechanics is the first system to possess certain fundamental properties which later came to be considered the hallmark of "scientificity."

We have already insisted on the importance of taking into account an entire system rather than a series of *ad hoc* laws, and we shall return to this point. Let us now consider the characteristics of Newton's mechanics which make it so much the paradigm of all empirical science that von Helmholtz, at the end of the last century, was able to state that no explanation in the natural sciences is clearly intelligible unless it is expressed in the terms of the principles of Newtonian mechanics.

This is not the place to propose a detailed analysis of the principles of Newton's mechanics. We shall limit ourselves to those aspects of the system represented in the *Principia* that have direct bearing on our epistemological analysis.

Let us begin with the principle of inertia. Historians of science are divided on the issue of whether the "discovery" of this principle is to be attributed to Galileo or to Descartes. A third possibility exists, which is a compromise solution. It requires us to accept the following proposals: (a) "Galileo did not have the modern conception of inertial motion"[1] as uniform speed in a straight line because "according to Galileo's principle of inertia, if the surface of the earth were perfectly smooth, a spherical body set in motion on that surface should continue to roll round the earth indefinitely";[2] (b) "Descartes improved upon Galileo by suggesting that natural motions took the form of uniform speed in a straight line, not a circle as Galileo had supposed."[3]

One might conclude from this analysis that Newton added almost nothing new to Descartes' conception. Even such an authority in this matter as Alexandre Koyré came to defend this point of view: "The classical conception—Galilean, Cartesian, Newtonian—of motion appears to us today not only evident, but even "natural." And yet this "naturalness" is still quite recent: it is barely three centuries old. And we owe it to Descartes and Galileo."[4]

It is true that Korýé continues the passage, adding that "the principle of inertia did not appear ready made, like Athena out of Zeus' head, in Descartes' or Galileo's thinking. . . . The Galilean and Cartesian revolution—which remains a revolution for all that—had been prepared for a long time." Still, the transition from the notion of inertia in the Cartesian sense to Newton's law is neither evident nor direct. In this sense, the "long preparation" which led to the law of inertia does not culminate with Galileo and Descartes; rather, these two scientists are part of the long path that was to lead up to Newton.

As I. Bernard Cohen pointed out, Galileo's idea of inertia is far from being precise. In some of his writings he referred to a concept of a circular, inertial motion; in other texts he seems to say the opposite. In fact, one finds in his writings several references to the fact that, given the dimensions of the earth, a horizontal movement (which coincides strictly with a maximal arc of a circle) can be considered as rectilinear. By bringing together the various references to this problem in different parts of Galileo's writings, it is possible to extract the following three conclusions:

(a) there is insufficient evidence to decide whether Galileo conceptualized inertia as rectilinear or circular;
(b) it appears that Galileo gave attention only to movements of limited extension; this may explain the ambiguity noted in (a);
(c) nowhere in these texts is there any reference to an inertial motion outside of the influence of the earth.

From this it can be concluded, as Bernard Cohen rightly points out, that several transformations were necessary to get from the Galilean concepts, via Descartes, to the Newtonian concept of inertia. According to Cohen, the following transformations were necessary:

(1) to establish with precision that inertia is exclusively rectilinear;
(2) to extend the consideration of movements which occur on planes near the surface of the earth to all rectilinear movements ("unsupported" by any plane);
(3) to imagine that the real world of inertial motions can continue indefinitely;
(4) to establish as an axiom that inertial motion is a "state";
(5) to associate inertia—and consequently inertial motion—with mass conceived as the quantity of matter.

Cohen is of the opinion that both Newton and Descartes contributed to the second and the fourth of these five "transformations," and that the fifth was Newton's original contribution.

At this point Cohen's historical analysis ends. We shall start here with our epistemological analysis. Cohen's points 4 and 5 are of particular relevance for this epistemological analysis. They indicate in what sense it is necessary to attribute to Newton, and not to Descartes, the merit of having established the concept of inertia on foundations that were sufficiently solid to become the basis of "classical mechanics."

The difference between medieval Aristotelian and classical mechanics has often been portrayed as a difference in the conception of motion: for the former, motion is a *state.* In fact, the Aristotelian tradition required the permanent presence of a *motor,* without which it was impossible to account for motion. However, Dijksterhuis pointed out that this difference can apply only to rectilinear and uniform motion, since in all other cases both doctrines require the presence of a "motor."

To say that rectilinear, uniform motion constitutes a process and not a state is equivalent to saying that no motor needs to be present. Descartes already had this idea, but it did not become fruitful until Newton gave it a precise quantitative sense and a relational interpretation. On this point, we diverge from Dijksterhuis (with whom we are otherwise in perfect agreement). Let us review Dijksterhuis' interpretations. His point of view is the following: defending or rejecting the thesis according to which inertial motion is not a process depends entirely on one's philosophical stance. He expresses this as follows:

> When he believes in the existence of an Absolute Space and an Absolute Motion in the Newtonian sense, regarding the latter as change of absolute place, when then, on the ground of the causality principle, he inquires after the cause of this change, and assigns Quantity of Motion as such, he will not have the least objection to setting up a very close relationship between Quantity of Motion and *impetus,* unless he prefers to consider the cause of persistence in uniform rectilinear motion to be inertia. In the latter case he may even appeal to Newton as his authority, for the latter ascribes inertial motion explicitly to a *Vis Inertiae,* Force of Inertia, residing in the body.[5]

Dijksterhuis concludes from this that even Newton may be considered to belong to those who conceive motion as a process. If this were true, one would have to establish a new distinction between Newtonian (which owes much to medieval mechanics) and classical mechanics as we know it today.

We believe that Dijksterhuis' statement is justified, but only if one considers exclusively Definition III of the *Principia:* "The innate force of matter (*vis insita*) is a power for resisting, by which every body as much as in it lies, continues in its present state, whether it be of rest, or of moving uniformly in a right line." (Translation by Andrew Motte, in F. Cajori, ed., *Sir Isaac Newton's Mathematical Principles of Natural Philosophy* New York: Greenwood Press, 1902). The problem is that this definition does not correspond to the law of inertia. When Newton formulated that law, he did not use the notion of *vis inertiae:* "Every body continues in its state of rest or of uniform motion in a right line unless it is compelled to change by forces impressed upon it." (*ibid.*, p. 13). There is here not the slightest trace of "inertial movement-as-process."

Now, this law cannot be understood as an isolated statement. It is part of a set of three, which constitute Newton's initial set of axioms. Newton, an heir of Euclid, knew very well that an axiomatic system is precisely that, a system. The "Scholium" which follows the three laws and their six corollaries begins as follows: "The principles which I gave up to now are accepted by all mathematicians and have been confirmed by an infinite number of experiments. The first two laws of motion and the first two corollaries enabled Galileo to discover that the fall of bodies varied as the square of time and that projectiles follow a parabolic path" (*ibid.*, p. 21).

Is it possible to be uncertain about what Newton was getting at? He thought of his *"Axioms, or Laws of Motion"* (this is the title he gave to this particular part of his *Principia*) as fulfilling the same function as Euclid's in relation to all theorems his predecessors had demonstrated. For this reason, it is absolutely necessary to read the first law in the context of the other laws. In considering together the first two laws it becomes clear in what sense Newton transformed (according to I. B. Cohen's expression) the Cartesian conception. The *state* of motion (or rest or of uniform rectilinear motion) is characterized by a precise value of a well defined quantity: (*mv*). The second law specifies how this state gets modified.

This leads us to consider Newton's second law in greater detail.

Whoever reads the *Principia* for the first time experiences the same surprise: the laws as formulated there do not contain the formula $F = ma$, nor any of its variants in differential form, such as are found in the manuals of physics where Newton's formula is presented. In the English version, the law is formulated as follows: "The changes that occur in a movement are directly proportional to the moving force and continue in a straight line in the direction of the force impressed" (*ibid.*, p. 13). This formulation, with its comment that follows as a clarification to the law, had led historians to think that Newton was influenced by the laws of impulse and to a lesser extent that of "instantaneous force."[6] Such an interpretation totally ignores Definition IV in which Newton defined what he meant by *impressed force;* it also ignores the comment which follows the definition and where Newton stated explicitly "The impressed force can have several different origins. It can be produced by a shock, by pressure and by centipetal force" (*ibid.*, p. 2). (He adds gravity between the centripetal forces in the comment to Definition V.) Such an interpretation equally neglects to take into consideration the *use* Newton made of this law in solving a great number of problems in the three books of the *Principia.* Furthermore, Cohen's analysis shows conclusively which is the proper interpretation of the Newtonian text and how it is to be related to the more modern version of Newton's second law.

From all this it follows that even though the formula $F = ma$ (or one of its variants) does not appear in this form in the *Principia,* it still expresses adequately the content of Newton's second law as formulated originally (without being a direct translation of it). It is also clear that this second law referred to *impressed forces* of various origins (percussion, pressure, "centripetal" forces such as gravity, etc). This is the starting point for our epistemological interpretation of the second law.

1. The Definition of Mass

The most widespread interpretation, so far given to the significance of these Newtonian laws is Mach's. Following this interpretation, Newton's third law permits a definition of the notion of mass, given which the second law becomes simply a definition of force. Hence, it becomes important to ask how valid these two statements are.

Mach used the following procedure. Consider two particles, of mass m_1 and m_2 respectively (whose relationship is to be determined). These occupy, at a given moment, positions $P1$ and $P2$. Let us call a_{12} the acceleration produced on the particle of mass m_1 by the particle of mass m_2; and let us call a_{21} the acceleration produced on particle of mass m_2 by the particle of mass m_1. Given Newton's second law

$$\underset{(m_1)}{\overset{P_1\,\vec{a}_{12}}{\longrightarrow}} \quad \underset{(m_2)}{\overset{\vec{a}_{21}\,P_2}{\longleftarrow}} \quad m_1\,a_{12} = m_2 a_{21}$$

that is:

$$\frac{m_1}{m_2} = \frac{-a_{21}}{a_{12}}$$

Taking the particle of mass m_2 as a referent (that is as a unit of mass), then mass m_1 is determined completely by the accelerations of the two particles at a given instant. We shall designate m_{12} the number which represents the value of the mass of particle P_1 in relation to the mass of particle P_2—that is:

$$m_{12} = \frac{a_{21}}{a_{12}}$$

Hence, we can in principle accept Mach's interpretation that the relation between the accelerations of the two particles "defines" the mass of one of the particles in relation to that of the other. Nevertheless, nothing as yet entitles us to suppose that the mass thus "defined" is an intrinsic property of the particle. For this to be the case, we would have to be able to show that the value of the relation $m_{12} = m_1/m_2$ is constant for any system of particles that includes the respective masses. Let us examine what happens when there are more than two particles.

When there are more than two particles in the system, it is absolutely necessary to consider the vectorial character of the accelerations. Let us designate $\vec{a}_{ij}$ the vector acceleration which measures the magnitude and the direction of the acceleration produced by particle j on particle i. Then, we can consider this vector as the product of its absolute value a_{ij} and a vector-unit $\vec{u}_{ij}$ which determines its direction and magnitude in the following way:

$$\vec{a}_{ij} = a_{ij}\,\vec{u}_{ij}$$

This acceleration cannot be determined by observation, since the only thing "observable" is the *total* acceleration a_i, of the particle P_i, and not the partial acceleration produced on P_i by each of the n particles that constitute the system. Therefore we know that

$$\vec{a}_i = \vec{a}_{i1} + a_{i2} + \ldots \vec{a}_{in} (i \neq 1, 2, \ldots n)$$

In this equation, only the first member can be determined by observation. Consequently, we have n vectorial equations of this type, one for each particle. We can write this summarily:

$$\vec{a}_i = \sum_{j=1}^{n-1} a_{ij}\,\vec{u}_{ij}\ (i \neq j)(1)$$

What then does it mean to say that the relation between the mass of a particle P_i and a reference particle P_2 is determined by the relation between their respective accelerations, supposing that no other particle is present? It means simply that we shall be in a position to establish

$$\frac{m_1}{m_2} = \frac{a_{2i}}{a_{i2}}$$

But this is equivalent to determining a_{2i} (not observable) from the *total* accelerations (observable) a_2, a_i. In other words, this is equivalent to solving the system of equations, in which the a_{ij} are the unknown quantities and the a_i are the givens. The answer to the problem posed thus depends on the answer to the question of whether or not the system of equations (1) has a unique solution. The answer is as follows:[7]

For $n = 3$ there are, in general, two different equations for the two a_{ij}, corresponding to each of the three particles. This is a unique solution. Hence, the relation between the masses of the particles can be uniquely determined (if the three particles are not all located on a single straight line).

For $n = 4$ there are, in general, three different equations for the three a_{ij} corresponding to each particle. There is likewise a unique solution (if the four particles are not in the same plane).

For $n > 4$ the number of independent algebraic equations in system (1) is still three, at least for the $n - 1$ partial accelerations a_{ij} corresponding to each particle. The problem remains indeterminate and the relations between the masses of the particles thus can not be determined uniquely.

The preceding problem is modified if we consider an observer who determines the accelerations of the n particles of the system at different moments in time (as many different moments as she wishes).

The maximum number of independent equations that may be obtained for all a_{ij} (that are now variable from one moment to the next) corresponding to each P_1 that has become displaced in space can be obtained by means of an analysis similar to the one given above, as follows:[8]

I. For $n \leq 7$ one obtains a system of equations that is determined and the relations between the masses of the particles can be determined uniquely under the condition that one has a sufficient number of "observations" of the accelerations of each particle at different times.

II. For $n > 7$, the system of equations is indeterminate and the relations between the masses of the particles cannot be established even if the number of times at which one measures the accelerations is very large.

The attempts to uniquely determine the relations between the masses of a system of particles using the principle of action and reaction leads, as we have seen, to failure. Mach's interpretation, according to which mass is defined by Newton's third law in such a way that the second law can be interpreted as a definition of force, is thus unacceptable.

There remains, however, the possibility of simultaneously determining the values of the force and the relations between the masses of a given system of particles by taking at the same time the second and the third of Newton's laws as well as the consequences that we can derive from them directly.

The analysis of this problem[9] shows that, by making use of the equations which correspond to the conservation of the quantity of motion and to that of the moment of the quantity of motion, it is possible to find solutions for certain cases where $n > 7$. These solutions correspond to particular configurations of the system of particles. However, there is no solution for the general case, nor are there any known criteria for establishing the cases when a system is or is not susceptible of a unique solution.

The preceding analysis reveals that Newton's three laws as formulated are insufficient for completely characterizing the notion of mass and of force, which constitute the principal concepts referred to in these laws. Consequently, we have to introduce more precision into the definition of these concepts or else establish new relations which would enable us to characterize these notions uniquely.

Now, these two concepts are not equal in terms of the conditions they are subject to. In the case of mass, we can postulate an intrinsic property of particles, that is, a property whose numerical value for given particle, is the same whatever the system to which it belongs and whatever its position and its velocity. By adding as a postulate the *constancy of mass,* we can determine the mass of a particle in simple situations—that is, in situations in which the three laws cited above are sufficient to determine the relation between masses uniquely—and attribute the same value irrespective of the situation where the particle is found.

The situation is different with respect to the force exerted on a particle, since its value changes not only when the particle belongs to different systems, but also within the same system for different moments at different time intervals. Hence, a single postulate would not be sufficient to provide the required precision with respect to this concept.

Difficulties of this kind appear when one considers the somewhat nebulous manner in which the classical texts refer to the concept of force. Considering, for example, the analytical mechanics of Lagrange, who represents the culmination of Newtonian mechanics, we find the following reference to the concept of force:

> By force or power one understands in general the cause whatever it is that impresses or tends to impress movement to the body to which it is assumed to be applied; and it is also by the quantity of motion imparted or which it is prepared to impart that one can get an estimate of the force or power. In a state of equilibrium, the power is not actually exerted; it only produces a tendency for motion; but it must always be measured

by the effect it would produce if motion were not arrested. In taking any force whatever or its effect as a unit, then any other force can be expressed as a ratio, a mathematical quantity that can be represented by a numbers or by lines; this is the way one should consider forces in Mechanics.[10]

When Lagrange defines the notion of dynamics, he gives somewhat greater precision to his concept of force:

Dynamics is the science of the accelerating or decelerating forces and of the various movements they should produce.[11]

Somewhat further on he adds:

We shall consider here mainly the accelerating and the decelerating forces, whose action is continuous, such as that of gravity. These tend to impress, at each moment, an infinitely small speed of equal magnitude upon all material particles. When these forces act freely and uniformly, they necessarily produce speeds that increase in time. One can regard the velocities thus produced in a given time as the simplest effects of these kinds of force, and hence as the most appropriate effects by which they can be measured. In Mechanics, it is necessary to take the simple effects of forces as known. The art of this science consists uniquely in being able to deduce compound effects that necessarily result from the combined and modified action of the same forces.[12]

The key expression in the preceding citation is to be found in the statement: "In Mechanics, it is necessary to take the simple effects of forces as known." In fact, the development of mechanics requires that the "simple effects" of forces be known. In other words, one has to identify the different types of force, or, put differently, the different types of systems of particles, where each is characterized by specific relations between particles. These relations make it possible to postulate the presence of a certain "simple effect" or of a particular type of force. Lagrange himself offers a first classification of the different types of systems.

All the systems of bodies which act upon each other and whose movements can be determined by the laws of Mechanics can be divided into three classes. For their mutual action can take place in but three different ways that are known to us: either through the force of attraction, when the bodies are in isolation, or through the ties that unite them, or finally through immediate collision.[13]

The first classification—even though it turned out to be insufficient—makes it possible to group the systems of particles into different classes, where each is characterized in terms of the domain of application of different forces. In each of these domains, a specific law (or a characteristic principle) applies, and this law or principle is valid for particular situations of a certain class of systems of particles.

2. The Domain of Elastic Force

When two bodies P_1 and P_2 are linked together by a spring (with certain characteristics), the force exerted upon them when they are drawn apart from the position of equilibrium obeys Hooke's law, which can be formulated as follows:

Let d_{12} be the distance of "equilibrium" between the spring and the two masses P_1 and P_2. If one, then, moves these two masses

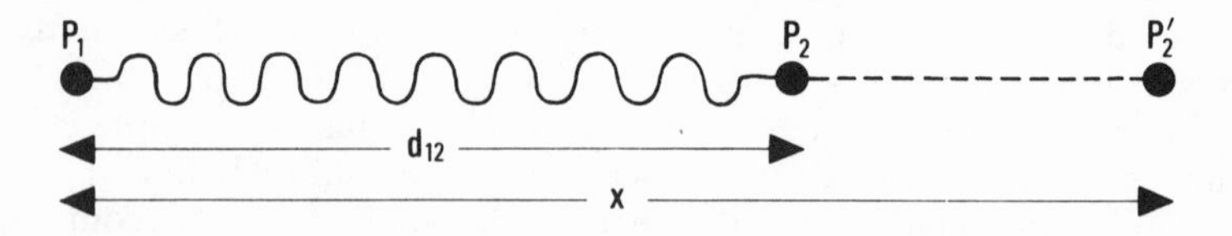

apart until they are separated by a distance x, the force exerted on them has the magnitude:

$$F_{12} = k_{12}(x_{12} - d_{12})$$

where k_{12} is a constant which depends only on the characteristics of the wire.

Let us further assume that the mass m_1 is linked to other masses m_2 and m_3 by means of the same type of spring (that is, one that obeys Hooke's law). The force acting on m_1 when the situation of equilibrium is destroyed will impress on it an acceleration a_1 in such a way that by using conjunctively Hooke's and Newton's second law we can write:

$$m_1\vec{a}_1 = k_{12}\vec{r}_{12} + k_{13}\vec{r}_{13}$$

where the vectors $\vec{r}_{ij}$ have for magnitudes the distances $(x_{ij} - d_{ij})$.

To the extent that the equation just given is valid for each particular position of the three masses, we can repeat the experiment three times, each time measuring $\vec{a}_1$, $\vec{r}_{12}$ and $\vec{r}_{13}$. We will then have a system of three equations with three unknowns (m_1, k_{12} and k_{13}), which makes it possible to compute m_1.

In a similar way, it is possible to compute m_2 and m_3 and, in general, any number of masses linked together by a spring.

Consequently, Hooke's law determines a new type of system which constitutes its domain of application.

3. Interrelations Between Particular Laws

We have shown that Newton's three general laws do not permit a characterization of the concepts of mass and force, to which the laws make reference, in a unique way. On the other hand, in the preceding section we have given examples of the way to proceed with systems for which certain particular laws are valid. Such examples could be multiplied by citing cases like scales, Atwood's machines, and the pendulum of Cavendish. In each case, a specific problem is solved by introducing a certain type of supposition or by using certain special laws. The resulting construction is nevertheless very coherent, which entitles us to consider all the systems involved in this construction as the domain of application of general laws of mechanics. What, then, is this generalization based on? How do we justify speaking of the *same mass* with respect to a "particle" whether it is determined by means of springs, scales, or the law of gravitation?

4. The Structure of the Classical Mechanics of Particles

We have already indicated the variety of types of "systems of particles" with respect to which the laws of Newtonian mechanics are valid. These kinds of systems are characterized as domains of application of special laws. Now, Newtonian mechanics is not constructed by simple juxtaposition of these domains. What is fundamental is the coherence of the total construction at which one arrives and which entitles us to consider all these domains as subdomains of the domain of application of the general laws of Newtonian mechanics.

This diversity within unity accounts for the considerable degree of confusion which has persisted during the entire history of mechanics until quite recently with respect to *the* role of Newton's laws (in particular, the second). Near the end of the last century, Poincaré tried to summarize the situation: "The principles of dynamics first appeared to us as experimental truths; but we have been obliged to use them as definitions."[14] Poincaré's formulation shows an important difference in conception with respect to Mach and Hertz, while not adding any clarification to that given by those authors. On the other hand, we have shown that Newton's general laws do not allow us to determine either mass or force except in very specific cases. It follows that neither Mach, Hertz, nor Poincaré was able to explain the functioning of the laws of dynamics.

It was not until very recently that the "eclectic" position found explicit acceptance. According to this position, the law $F = m(d^2s/dt^2)$ has several distinct uses in mechanics. Russell Hanson has provided what is probably the clearest analysis of this situation. He has identified at least five "different uses" of the celebrated law of Newton. After having enumerated these, he states:

> The actual uses of $F = m(d^2s/dt^2)$ will support each of these accounts. This means not just that among physicists there have been spokesmen for each of these interpretations, but that a particular physicist on a single day in the laboratory may use the sentence $F = m(d^2s/dt^2)$ in all the ways above, from 1–5, without the slightest inconsistency.[15]

Nevertheless, recognizing the multiplicity of the functions of the law cannot in itself explain the structure of Newtonian mechanics. Hanson's penetrating analysis as well as those of other philosophers of science who have recently written on this subject, in very similar terms, serves to formulate the problem but not to solve it.

From our point of view, taking as a basis the analysis presented above, the problem may be formulated as follows.

Each of the particular laws which we found to be applicable to a specific type of system has its own domain of application. We cannot apply springs to the planets, nor can we put them on scales. Neither can we apply the law of gravitation to any two objects

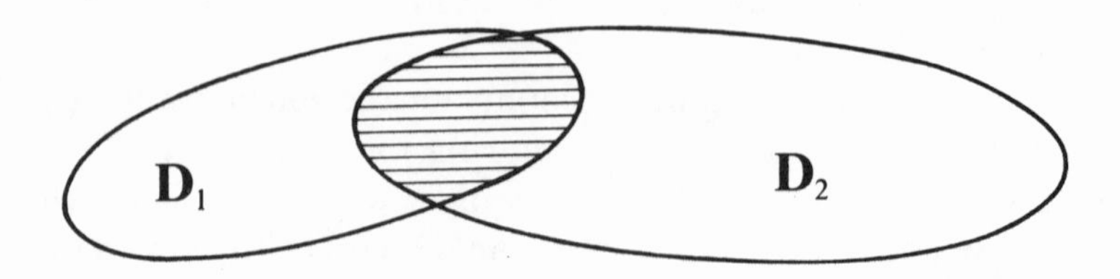

found in a laboratory, except under very particular conditions. But none of these "domains" is isolated. They all have "regions" where they appear as overlapping. Let us take an example:

Let us suppose that D_1 is the domain of application of Hooke's law (that is, in D_1 are to be found all the systems of bodies that are subordinated to elastic forces of a certain type). Let us further assume that D_2 is the domain of application of the law of gravitation (our planetary system, for example, belongs to this domain). The region common to the two domains represents all those systems in which both laws apply in particular situations. A body suspended by a spring near the surface of the Earth allows us to study the oscillations of the body and to show that this movement is in accordance with *both laws* (Hooke's law and the law of gravitation). Hence, it is possible to identify the masses one might have obtained by applying each of the two laws separately.

These considerations lead us to conceive of Newtonian mechanics as a complex structure made up of different overlapping domains, characterized by *specific laws* or principles. All the domains, however, can be subsumed within one global domain, in which Newton's three laws are valid.

(a). Mass and force, which are *involved* in the second law, are theoretical functions which correspond neither to a universal definition nor to a single method of determination. Only in particular domains of application can masses be determined by an appropriate method. Similarly, there is also in the case of force a particular law that is valid for each domain (the law of gravitation, Hooke's law, etc.).

(b). Mass is considered as an intrinsic property of the particle, which remains invariant for any domain of application of the theory. It is as a function of the structure of the second law that mass receives its property of parameter intrinsic to the particle. Hence it is *the law itself which relates the different domains* in a single structure with regard to the relationship between mass, force, and acceleration. The "synthesis" of the different domains in a single structure is accomplished by the theory. Therefore, the hypotheses concerning the different forms the theoretical functions take in the different domains are not independent with respect to each other.

II. EPISTEMOLOGICAL REFLECTIONS ON THE EVOLUTION OF MECHANICS

1. Observables, Theoretical Terms and Theories

In order to formulate the conclusions we can draw from the historical development of mechanics, to which we have made reference in this chapter, we shall consider separately the two aspects, which taken together, constitute a physical theory: (1) the structure of the theory itself; (2) the elements upon which this structure operates and the kind of "correspondence" that may exist between these elements and physical experience. This distinction is, to be sure, only a methodological one and serves only to underline certain aspects which have certain characteristic properties. The theory as an explanatory system is an integrated unit, where these aspects fulfill certain functions that obey the laws of the total structure. To accept a physical theory is to suppose relating a "theoretical framework" (that is, a certain formal structure) to a set of "objective situations" (that is, a set of objects and their interrelations). The empirical content of a theory can be expressed by saying that the theoretical framework within which the theory is formulated is *applicable* in a given situation. We can express this differently by saying that the theoretical framework is an adequate model of the situation. We are using the term model here in a totally different sense from that used in logic and mathematics, where the term "model" designates a precise interpretation of the undefined terms of an abstract axiomatic system. A physical theory constitutes an axiomatic system that is already interpreted. This abstract system together with its interpretation, is what we call a "model" of physical reality. (Or, more precisely, of the domain of physical reality which we are trying to explain.) To "explain" in physics, then, is to formulate an adequate model of a group of phenomena. Now, two kinds of problems arise immediately: (1) Which facts are to be taken as a starting point to define or describe the phenomena to be explained? (2) What are the characteristics of the models which have been accepted historically as being "adequate" and how has the construction of these models been accomplished?

With respect to the first point, the analysis we have made of the construction of theories, both in classical and in quantum mechanics, leads us to formulate the following observations:

a. Each theory corresponds to a certain level of "abstraction" with respect to physical reality. At each level certain properties of the objects to which the theory applies are taken as a point of departure. These properties constitute the "observables" *for the theory in question.*

b. The characterization of the observables poses no philosophical or metaphysical problem as such; neither does it presuppose the acceptance of an irreducible *a priori* for the theory in question. The theory limits itself to recognizing that certain variables may have values which can be obtained by procedures that are justified outside of the theory. For example, in classical mechanics we have taken as a point of departure the fact that each particle has a position in space, a position that is well defined at every instant of time, and a certain velocity, also defined for each instant in time. Classical mechanics as such does not raise the problem of how to obtain the "position" of each particle or the interpretation given to the concept of time used to describe the "successive" positions of the particle.

c. In the description of the phenomena to which the theory applies, one also uses other concepts which are not taken as data, since the theory itself will take on the task of characterizing them with precision. These concepts correspond to "functions" (in the mathematical sense of the term) whose values can be obtained only by applying the theory to the objects that belong to a certain domain at the level of the phenomena one wishes to explain. These concepts are usually called "theoretical" terms.

Two consequences follow from the preceding. First, the classification of the terms of a theory into what is "observable" and what is "not directly observable" or, more precisely, into "nontheoretical" and "theoretical" is specific to each theory and has no meaning outside of the context of that theory. But, on the other hand, nontheoretical terms (the "observables") for a given theory are the product of prior theoretical constructions. For example, space and time, which are part of the definition of the position and the velocity of a particle in classical mechanics, are complex theoretical constructions. The theory that accounts for the use of these constructs is based on other "observables," with respect to which, within this theory, the position and the velocity of the particle at a given instant become theoretical concepts.

2. "Empirical" and "Reflective" Abstraction[16]

Starting with the traditional concepts accepted for the construction of classical mechanics, it is thus possible, on the basis of what we just said, to take a step backwards in the construction of concepts. In this way, we shall get to the genesis of the basic concepts which are part of the construction of the outside world. This construction originates in the subject's actions. But we can also proceed in the opposite direction toward increasingly higher levels, up to the level of the most complex theories of modern physics. Thus, we can reconstruct the successive processes, beginning from the "conceptions" of an infant who does not yet crawl on hands and feet, up to those of physicists, who constantly search for new particles with strange characteristics to "explain" certain "incomprehensible" phenomena.

The point we are trying to defend in our work is that these processes have the same characteristics from one end to the other of this scale. We can speak of a single process divided into phases. At each level, certain previous constructions remain in acceptance while other new constructions become elaborated. This is true both for the child and for the quantum physicist. What is characteristic of the process is that at each new level, there is a return to the "level of experience." Each new level is equipped with new interpretative schemata, which enrich the original notions used in the construction of that level. But this enrichment does not consist only of the discovery of new properties of objects or of new relationships between objects. Very often, it is the object itself that gets modified, and this modification has a very precise meaning: the fact is that certain initial properties of objects can no longer be accepted, or else they lead to contradictions within the interpretative schema. These are the properties which we are *forced* to give up to leave intact the structures which make the whole experience intelligible.

By "intelligible" we refer to the fact that, when one makes use of the relations established by the theory, certain results of measurements come to appear as necessary consequences of other measures. This is the case when the same theory can be applied to many different systems.

These characteristics of physical theories (their capacity to interrelate the values of nontheoretical functions which belong to various domains of application, and their return to the level of experience and the modification of basic concepts) show that we are *not dealing here with a sequence of structures that are mutually subsumed by one another.* The process is much more complex. This complexity is reflected in the difficulties inherent in all attempts at formalization.

Again, in spite of the complexity of the process, there is pervasive regularity and uniformity in the methods of construction. No matter what the level of abstraction to which we penetrate, the transition from one level to the next always comes about in the same way across the two types of abstraction. One begins with certain concepts (which are the product of the reflective abstractions of preceding levels) so that via a process of empirical abstraction, one is able to identify certain functions upon which to build new concepts that are directly applicable to a certain domain of reality. Then, by a process of reflective abstraction—which makes use of a mathematical framework identical to the structure of the theory being constructed—one extends these concepts to other domains of reality. In this way, these domains become, in turn, intelligible. These two processes of abstraction constitute the method of construction of all physical concepts and, at the same time, the link connecting the different levels of theoretical construction. Let us now formulate what we have just said with somewhat greater precision.

3. The Succession of Theories

The starting point of a physical theory T^j is always a phenomenon about which one seeks an explanation. The choice of this phenomenon (or group of phenomena) *partitions* a part of reality. That is, in the description of the phenomenon one considers (where necessary) certain individuals while ignoring others; similarly, one refers to certain relationships present between these individuals, while ignoring other possible relations. Therefore, from the very beginning there is a process of abstraction, which we designate A^i. It is by means of this process of abstraction A^i that we define a situation S (certain individuals and certain relations between them). This situation exhibits the phenomenon one seeks to explain (F).

The description of F presupposes, in general, certain concepts 0^i which correspond to "direct experiences" of the phenomenon F. Some of these concepts are only vaguely defined, even though they presuppose very complex levels of abstraction (for example, Newton's concept of mass conceptualized vaguely as quantity of matter). It is thus possible to analyze these concepts to characterize the process of abstraction which led to these concepts. Nevertheless, such an analysis is not necessary from the point of view of physical theory, since the theory itself will assume their definition and precise characterization.

The mechanism is the following:

a. There are certain concepts of type 0^i, which, unlike the others, constitute the *data* the theory accepts as being sufficiently well defined (or well characterized, or as already constituted). The concepts of space and time, with their respective measurement scales, as well as the concept of "position," which is defined by them, constitute examples valid for classical mechanics. In other words, from among the total set 0^i the theory accepts a subset 0^i_0 and reconstructs the others 0^i_1. This subdivision of the 0^i into 0^i_0 and 0^i_1 is always relative to a particular theory. For example, space and time, with their respective measurement scales, are 0^i_0 when the theory T^i is classical mechanics but they are 0^i_1 for relativity mechanics. We shall call *observables* in relation to a theory T^i, the 0^i_0 accepted by that theory. These observables are "drawn" from the plane of experience by a process of empirical abstraction, we shall indicate by A^i_0.

b. On the basis of the total set 0^i relative to a type of phenomena F, the theory thus begins by establishing a distinction between the 0_0^i and the 0_1^i. This process consists, in general, in a precise characterization of the 0_1^i within a certain domain of application of the theory T_1^i. The domain of application is defined by a subset of objects, events, etc. that are accepted as such in situations S, which constitute the point of departure. This domain is characterized by the fact that the 0_1^i are uniquely determined by the 0_0^i within the domain. We shall call process of abstraction A_0 *relative to a theory* T^i the process of construction of the 0_1^i on the basis of the 0_0^i within a particular domain.

c. A theory that stops there is of little interest. Theories (to the extent that we take them to be "explanatory") become interesting when they succeed in applying to other domains at least one of the 0_1^i that are completely characterized within a domain D_1^i. To be sure, in the new domains (D_2, D_3, etc.), the same 0_0^i remain valid. In these domains, the type of abstraction A_0^i is not applicable (if it were it would be the same domain D_1^i and not another one to arrive at 0_1^i). This takes a *theoretical construction.* The pro-

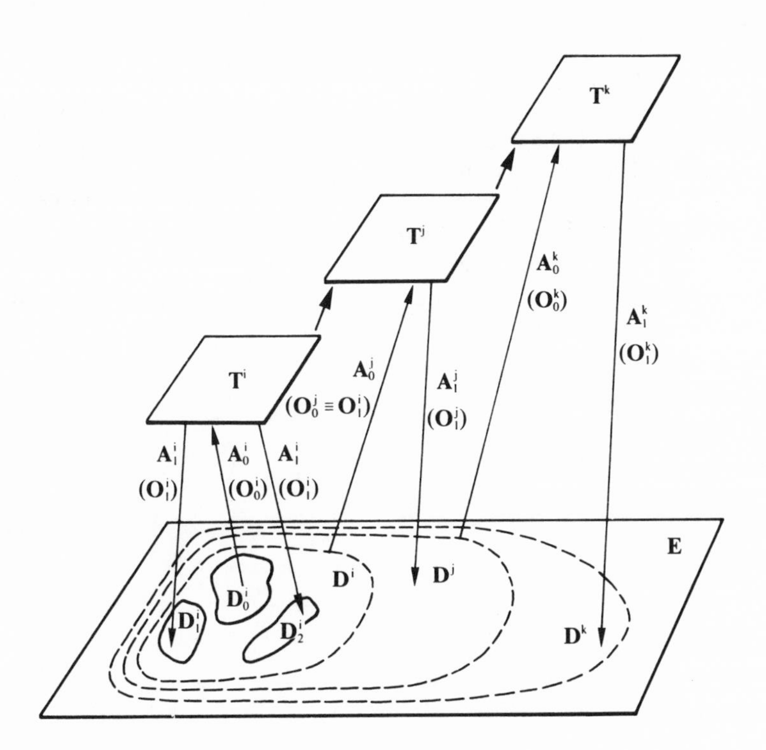

cess which leads to the construction of an 0_1^i in the domains D_1, D_2, . . . D_n will be called a process of reflecting abstraction A_1^i (to the extent that the new 0_1^i is *reducible* to the old one when the theoretical construction is applicable to D_1^i, since only in that case can we continue to speak of 0_1^i).

d. The transition from a theory T^1 to another T^j is such that: (1) The 0_1^i of T^i (or at least some of them) are considered as 0_0^i in relation to T^j; (2) T^j is applicable to all the domains D_i^0 while also comprising other domains (that is, $D^i \subset D^j$). The result is that the process of abstraction which has led to the 0_1^i (abstraction of level A_1 in relation to T^i) is now considered as A_0 in relation to a new theory.

e. The A_0 of a theory are *empirical type abstractions* in relation to the 0_0^i that have served to characterize the F_0 of this theory. The A_1 are *reflective abstractions.* The *product* of a reflecting abstraction in relation to a theory becomes the *observable base* (the observable starting point) for an empirical abstraction in relation to a higher order theory.

D = Domain
E = Plane of experience, within which are given S (the situation) and F (the phenomenon). It is the locus of the "observables."
T^i, T^j, T^k = The successive mechanical theories.
A_0^i, A_0^j, A_0^k = Empirical type abstractions in relation to the theories.
A_1^i, A_1^j, A_1^k = Reflecting abstractions in relation to the theories.
0_0^i = "Observable" in relation to T^i.
0_1^1 = Theoretical construction in relation to T^i but "observable" in relation to T^j.

VIII

The Psychogenesis of Physical Knowledge

I. INTRODUCTION

In physics, we find ourselves facing problems of construction and validation of knowledge that are far more complex than is the case in disciplines with an endogenous source. Since the principal necessary condition of truth for a physical theory is its adequacy with respect to "facts" external to the subject, nothing could guarantee that we would find any regularities or general aspects in the construction of such theories. Since the times of Galileo and Newton, there has been an immense diversity of facts to be interpreted and, correlatively, a great variety of theories has been elaborated since physics became "scientific." While it is true that even at that time the only facts considered were those that were held to be measurable to various degrees, and while it is also true that a second necessary characteristic of physical theories is that they include a coherent form of mathematics, a considerable difference separates the sequential generation of logico-mathematical structures from that of physical theories. The structures were generated one from the other and those new structures either did not contradict or had to discard what had been demonstrated by the preceding ones. In contrast, one does not find such general integration in the history of physical interpretations. We do find certain trends that are more or less continuous, but also contradictions, corrections, and much trial-and-error behavior. Naturally, facts are attained only by successive approximations, and even sometimes with the necessity to abandon a property that seemed at first constitutive, but which, if maintained, would have been in contradiction with new facts. It was thus reasonable to ask if our search for common mechanisms of formation or transition would still make acceptable sense with respect to scientific physics or whether this search for relations with psychogenesis would not be entirely futile. While it is true that in chapter 1 we saw remarkable parallels in the history of the impetus in physics and the idea of momentum ("élan") in psychogenesis that seemed quite promising, these were only prescientific notions, which occurred before experiment, measurement, and the setting up of equations. Hence, we were quite uncertain about possible common mechanisms in the formation of the child's physical interpretations and theoretical processes at higher levels.

Now, historical-critical analysis provides a fairly enlightening answer to the question concerning mechanisms of formation and transition on the level of theories. It is enlightening through its combination of analogies and differences compared to what we found with respect to logico-mathematical structures. The first common feature—and it is fundamental—is that in considering the milestones in the development of mechanics, one encounters the same sequence of intra, inter, and trans levels: first, the passage from a situation of the intra type, where the data are subsumed under general laws that do not, however, generate these data and their differentiations; then, the inter type, where invariants are deduced from transformations; finally, the trans type, which continues with the subordination of the physical system under a general algebra whose new variables correspond to its operators. Furthermore, in certain cases, such as the one where Hamilton's functions were to be translated into quanta, one finds a second partial analogy with the way mathematical structures are generated: this is the necessity to pass again through the stages of classical mechanics and to equip these with their new quantum content. Hence, in spite of what appears to be the prototype of a Bachelardian "cut," comes the necessity for support from the "historical process of the formation of basic concepts" with their sequential embeddings.

Still, these analogies with the mechanisms of transition involved in the elaboration of mathematical structures are quite understandable, since mechanical theories have brought the interpretation of physical phenomena to their highest degree of mathematisation, owing to subjects and operations. However, differences necessarily appear with regard to the relations to be established between theories and facts, since such relations are valid only to the extent that they agree with experience. Now, experience tends to become enriched and modified continuously, and there does not seem to be any law regulating or predicting the sequence of discoveries, *a priori.* As a result, the differences compared to the lawful succession of mathematical structures may seem drastic while it appears that the general trends in the development of mechanics show various stages such that Newtonian mechanics represents what we may call an "intrafactual" stage, which was followed by an "interfactual" stage represented by Lagrange and Hamilton, and finally by a "transfactual" stage represented by algebraic microphysics.

It seems more difficult to subdivide these main periods into substages, where we would find a regular succession of intra, inter, and trans type of characteristics according to some intrinsic necessity which we would describe in terms of a process of equilibration. In other words, disequilibria and re-equilibrations follow each other in a more irregular fashion, which is all the more pronounced as the experimental discoveries are less well coordinated with each other. Hence it is all the more remarkable to be able to show, in spite of the presence of factural contingencies, a general functional process mediating the transition from the main results of our psychogenetic and historical analyses, which indicate that the "facts," in spite of their obvious relationships to exogenous sources (both direct and indirect), cannot be established independently of endogenous structures, by means of which subjects interpret the facts. These interpretations constantly mix factual observations and inferences (where the former may precede the later and vice versa).

Thus, at this level, where the subject interacts with the objects, we again find the usual process, with its succession of intrafactual, interfactual, and transfactual levels.[1] This time, however, they are found only in a very general, more elementary form. The intrafactual is a phase where a theory *Tn* is based on a certain number "of facts," obtained by simple empirical abstractions; it must, however, be understood (as we noted in our general introduction) that the latter presuppose the existence of assimilatory schemes that result from the subject's prior actions (thus involving an elementary form of reflective abstraction, which remains, however, instrumental only—that is, unthematized). In this case, the theory *Tn* unites under a general law, a certain number of facts, but without generating their differentiation (for example, in Newtonian mechanics, as well as in Euclidean intrafigural geometry, each problem requires its own method of solution). The phase leading to the following theory *T* (*n* + 1) is then characterized by constructions based on reflecting abstractions and completive generalizations, the former being realized within the system *T*(*n* + 1) in its formative phase (cf. the role of symmetry in Maxwell's equations) as well as deriving from *Tn;* on the other hand, the latter originate from endogenous requirements as well as from the integration of new facts. The interfactual phase thus represents the part played by the subject and her operations in the transcendence of theory *Tn* and the elaboration of new structures. Hence, the transfactual phase, where the theory *T*(*n* + 1) is subjected to verification by new experimentation, permits one to ascertain, by means of observables that can now be controlled through new empirical abstractions, what had remained purely inferential products in theory *Tn.* The term trans can be justified this way. The new observables in *T*(*n* + 1) constitute, together with those of *Tn*—to which they are added following their construction—another system that is larger again. Furthermore, differentiations characteristic of subsystems can be generated by the *structure d'ensemble* of *T*(n + 1), to the extent that the processes of the inter phase are operative. The same mechanism of formative interaction between the subject and the objects then repeats itself going from *T*(*n* + 2) to *T*(*n* + 3), etc., but with progressions that reinforce simultaneously, in the course of the successive triads, the internal necessity of models and the adequacy of the facts.

If such is the process of transition from one theory to the next, it is natural that its generality implies the assumption that its beginnings go all the way back to the initial stages of psychological development. In section III, this is what we shall try to show with respect to the notion of weight. But first, in section II, we shall analyze some examples showing the early emergence of reflective abstractions and of constructive generalization that are not purely extensional.

II. ABSTRACTIONS AND GENERALIZATIONS NECESSARY TO THE CONSTITUTION OF ELEMENTARY PHYSICAL FACTS

The development of empirical abstraction (abstracting from objects) and that of reflective abstractions[2] (proceeding from the subject based on the subject's actions and operations) are far from being parallel or symmetric: While the latter tends to liberate itself from all factual verification and to evolve toward "pure" logical and mathematical truths, the former increases in precision and efficiency, but only insofar as it receives from the latter the instruments of observation and elaboration. This happens as early as the most elementary reading of factual experience, while the subordination of factual to reflective abstraction intensifies in the course of development. It thus seems interesting, for the epistemology of physics, to seek to observe these relations from the start, that is, from the first constitution of "facts" (or observables interpreted, see our Introduction). Even such observables, as direct and uninterpreted they may appear to the subject, already imply the use of assimilatory schemata, which are indispensable to any "reading of facts" in as much as such reading is immediately interpretative.

We shall thus try to show that in the earliest levels of representation one already finds the six principal characteristics which our historical analyses have indicated in relation to high-level physical theories. These characteristics turn out to be the result of very general mechanisms of abstraction and generalization rather than of the growing complexity of problems and levels of knowledge. In fact, we shall note that in the course of psychological development (1) A single concept (for example, a pressure exerted by the weight of an object when placed on another object) may in some cases correspond to an observable (creating a depression, etc.), while, on other occasions, it corresponds to an inferential coordination without possibility of verification by the subject (hence its functional but not structural, equivalence with a "theoretical term" in a physical system). (2) The result is that, in the constitution of a field of knowledge, empirical abstractions characteristic of the reading of observables or of experimental facts and reflective conceptualized abstractions necessary for the construction of "inferred" facts or concepts inherent in deductive coordinations alternate with each other. (3) A single general relationship such as that of velocity $v = d/t$ can have a number of mutually independent domains or subdomains of application, which are not reducible one to the other, although there may be overlaps or intersections. There may also be subordination to a more general relation, but this remains global and does not generate these subdomains as differentiations. (4) Hence there is alternation between abstraction and generalisation, but (5) these appear in constructive and complete as well as in inductive forms, both at the level of physical and logical-mathematical knowledge. (6) They are frequently accompanied by a reinterpretation of variables within an extended set of principles of conservation.

1. Pressure[3]

It will be helpful in this regard to analyze some facts. One of the most instructive has to do with the young child's notion of pressure and resistance (or somewhat later his idea of reaction). In both of these notions the role of inferential coordination is considerable. At an initial level, the property or activity of "pressing" is reserved exclusively for those cases where a perceptual observable is involved, as in the special case of a depression (for example when a metallic cylinder is placed on a mossy surface). In such situations, subjects appeal to the weight of the agent and the softness (or the lack of hardness) of the reactant, both of which they consider as necessary and sufficient conditions. Of course, they do not formulate this notion explicitly, except in the special, negative case of "steel" on steel or of moss on moss. In these cases, subjects think that there is neither pressure nor resistance involved, since "they are both the same." As for steel on wood or on any other stiff surface (all situations are verified following a prediction so that nothing remains at the verbal level), subjects do not believe that there is either pressure or resistance, since no visible depression results.

The first phase, characterized by the presence of a single positive or negative observable (but one which is already a "fact" in that it receives an interpretation), raises in itself two problems: that of the conditions determining the reading of this observable, of which the subject thus becomes conscious; and that of the relationships, which are also important, although not given direct consideration. They instead await demonstration by reflective abstractions in the following period. As for the determining conditions, the comparisons and classifications carried out by the subject on the various instances of depression are logical or prelogical coordinations. They become operative as early as in sensorimotor schemata, in the absence of which the simplest observations could not be apprehended. Thus, these instruments of establishing connections are formed before actual behavior and become integrated in it. In addition, other relations come into play. Their role is still implicit but quite important: for example, when a subject predicts that a particular weight will produce a depression in the moss, she is sure that this will be true *a fortiori* for an object of superior weight. However, she is not yet conscious of this gradation, at least not to the extent that she expresses it verbally. At the following stage, however (between five and seven and a half years), subjects come to use a certain form of quantification: for example, they may say spontaneously things like the steel pushes "very hard" on the moss, "not very hard" on the sagex, and "a little bit" on a thin sheet of wood (Lau at 5;6 years).

Now, this first form of quantification, arising from relations implicit in the coordinations of level I through reflective abstraction, constitutes the source of an important generalization: if it is true that a given weight exerts different degrees of pressure depending on the degree of resistance offered by a reactant, then it becomes possible to seriate these effects. This means that the weakest pressure can be considered as positive rather than as null. We observed a clear case of this reasoning (Eri 7;2), but since he still relates pressure to depressions produced, he believes he is seeing them everywhere, thus constructing false observables (and he is not alone in doing so): "All this stuff, I can see it getting pushed down a tiny bit."[4]

The following phase in the development of these generalizations based on reflective abstractions brings about the following inference: Given that the different agents produce differential effects on the reactants, which vary in softness or "hardness" from one object to the other but not within a single object, it is resistance which thus comes to be distinguished from hardness as a simple static property; further, by extension of this recognition, one has to assume that each agent (steel, wood, etc.) also possesses its particular weight so that it should exert a certain uniform pressure. Hence, the variations observed are to be attributed only to the multiple combinations possible between agents of different weight and reactants of various hardness levels.

Thus, we have the beginning of the notion of pressure which can be generalized independent of the depressions produced. This notion operates even in cases where no verification is possible.[5] That is, it is no longer an observable in cases where the "reactant" is so "hard" that it comes to be promoted to the rank of an inferred concept and the generality of which becomes to be taken as necessary (about seven to nine or ten years, with spontaneous coordinations at eleven to twelve years).

Corresponding to this evolution of the notion of pressure we find a similar one with respect to that of resistance, in the sense of a "reaction" that is symmetric but delayed with respect to that of "actions," as we just described. At first, resistance is conceived only as a nonvectorial obstacle due to "hardness"; thus, it is not seen as a form of action. It is seen to vary from one material to the next, remaining constant for a given object. From seven or eight years on, resistance is seen to vary with the agents. Thus it comes to be perceived as a semi-activity. We find, therefore, a first emergence of reflective abstraction derived from the pressure variations recognized during the previous phase (five to seven and a half years). It is important to note that the relation is first thought to be an inverse relation: to a small amount of pressure there corresponds a large amount of retention and to a strong pressure a small resistance; thus, Dra (7;11) remarks, "the moss holds back less if the weight is heavier because it makes a deeper hollow," and this opinion can be found up to the age of nine or ten years. In other words, while it is true that evaluating hardness and softness requires only empirical abstraction whereas generalizations leading from the observation of variations to an understanding of the action of resistance presuppose a level of abstraction and quantification going beyond mere observations of facts, subjects still remain dependent on the depressions produced, as these are the only observables permitting them to relate pressures to resistance. Reflective abstraction becomes operative only from the moment when the child, instead of concentrating exclusively on observable results of certain pressure variations, begains to compare them as antagonistic actions. Only then will she begin to see them as symmetrical. Thus, Fré, aged 10;5 years, said: "It pushes harder, so the moss has to put up more resistance."

But three sorts of advances are still necessary before the idea of "reaction" gets developed. The first consists in substituting for this semi-action of "resisting," which is not yet vectorial in nature, a directed action which the subject expresses with the term "repel," which comes to be more and more equivalent to "push upwards" in response to the pressure exerted downward. The second advance, which is more difficult to accomplish, is to generalize this upward push to those cases where the agent's pressure is not observable and where, consequently, the repelling action of the reactant is even less so. The third development—not accomplished before eleven or twelve years—occurs when the subject concludes that there must necessarily be equivalence between action and reaction; otherwise one produces a visible effect on the other. Toi, at 11;2 years, went so far as to affirm that if the chair did not repel him he would push it down into the ground and if the pressure from the chair were not equal to that he exerted on it, the chair would propel him to the ceiling. Here one can see the indispensable role of reflective abstraction and constructive generalization, since the three advances come from inferences going largely beyond empirical observations. This does not mean that subjects do not also revert to verification on observables, as we noted in other situations.

These facts about pressure constitute a good example of the importance of the "inter" phase (described at the end of section I) in the change from one interpretative mode to the next, and they clearly show the complexity of this process. In fact, it is clear that after the "intra" phase, where only those instances of pressure which produce visible effects, such as depressions, are recognized as such, subjects generalize to cases where no perceptual cues are present and no empirical verification is possible. Now, these are not simple extensional generalizations, since they are mediated by reflective abstractions resulting from two fundamental schemata of operational structures in the subject: quantification (in terms of more and less) and reciprocity. The former leads to the idea that if depressions can be of varying degrees there may also be nonvisible, but nonzero depressions; this then leads to the idea of relating this capacity for exerting pressure to the respective constant weight of the objects. As for reciprocity, it leads subjects first to subsitute the notion of action resistance for that of hardness resistance of the apparently passive objects. This gradually leads to the notion of equality between action and reaction, acquired in stepwise fashion. All this mental work accomplished during the "inter" phase is, thus, purely inferential; it shows clearly the role of reflective abstraction, which finally leads subjects, during the trans phase and still within their own experience, to a level where the intrinsic necessity of the model becomes evident to the subject, without a need to resort to new observables. These were already constructed deductively in the course of the "inter"-phase.

2. Velocity[6]

Among the many domains in which one single law may appear in many different forms—a situation which creates special difficulties for the elaboration of syntheses—the most noteworthy example is that of velocity. Since the definition or constitutive relation of velocity is $v = d/t$ or $v = n/t$ if n corresponds to homogeneous frequencies of occurrence, velocity always implicates time, and even duration, as time intervals. But the generalization of this common condition is a late development. In fact, the earliest intuitions of speed even seem to lack it altogether—that is, they make use of the notion of temporal order, but not of duration.

1. In fact, the initial notion of speed is based on the process of one mobile passing another: if one of two mobiles (it takes at least two) is behind the other one at some point in time and then, at some later moment, is ahead of the other, then it is considered to be faster. In such a case, estimating the speed of the mobile only requires reference to spatial and temporal order. However, it is clear that spatial intervals (the distance covered by the two mobiles) and temporal intervals (the synchronous durations of the motions compared) are already implicitly present, without the subject being conscious of it. The problem is to understand what kinds of abstractions are necessary in order for the subject to construct the notion of interval, and particularly, to understand the role of duration in these relationships.

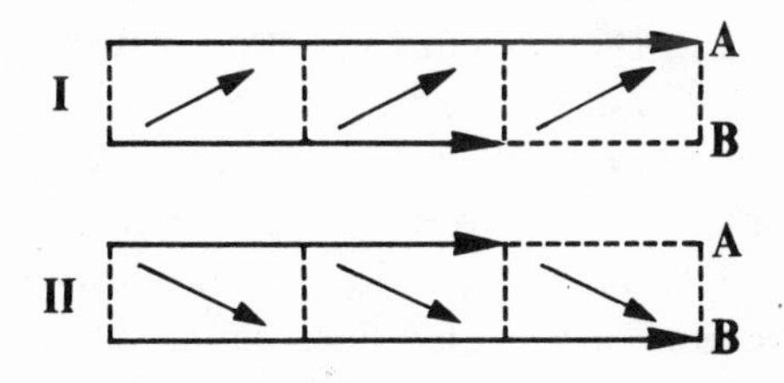

Let us begin, for this case, with the correspondences subjects in fact use when they correctly estimate the velocities or the durations on the basis of the point of arrival (to simplify we assume identical points of departure). The horizontal arrows in *I* represent spatial paths (→): mobile *A* is faster than *B* since it passes *B* and thus covers a greater distance. The oblique arrows represent in this case simultaneous events (↗), that is, the fact that, at any particular moment, *A* is ahead of *B*. In addition, we represent the internal references of the system by means of vertical dotted lines, which mark identical positions on the two parallel paths. As for II, it represents a different situation. Here, durations are evaluated by means of the order of times of arrival: the horizontal arrows (→) thus represent, in fact, durations. Mobile *A* is faster since it arrives before *B*. In this case, the oblique arrows (↗) connect identical positions, which *A* reaches sooner than *B* does. The vertical dotted lines mark the internal temporal reference system—that is, the simultaneous time intervals.

Before returning to our problem, let us note the complete symmetry of the two systems of correspondence, I and II, which illustrates once again not only that velocity implicates time but also that time is in turn inseparable from its content, that is, the speed with which events take place $t = d/v$. Young children, when they see a mobile arrive at the same spot before the other mobile, as in our illustration II, easily conclude that it moved faster and has used less time (the relation "less time" is sometimes expressed in the words "got there faster"). However, they frequently believe, in situations that require more interpretation, that the following relations hold: "faster = farther (*A* farther than *B*) = more time," because keeping speed constant, the relation "more distance = more time" holds. These, of course, are erroneous deductions, while as long as we stay with simple observables, as in illustrations I and II, the relations and correspondences children note are in fact those we described.

From these facts, it is then easy to see how the subject, basing her estimations of the ordinal relations involved only on the fact that one mobile passes the other, comes to conclude that $v = d/t$ taking account of the distances covered d and the durations t—that is, the spatial and temporal intervals rather than the final correspondences. Now, while the points of arrival can be perceived by simple empirical abstraction, the consideration of intervals requires use of reflective abstraction, since a retrospective and even a retroactive procedure of sorts is necessary in going from the terminal states to the antecedent events and the points of departure. Even where the latter differ—either spatially or temporally—for the two mobiles to be compared, young children tend to neglect this, as we saw above with respect to "commutability." To conceptualize intervals, one has to construct new relationships between the initial and the final points in space and in time, and to take motion as a continuous process in considering all of its successive states. Now, while it is true that all these aspects already play a role in subjects' actions, they do not necessarily constitute objects for thought. Rather they remain implicit and undifferentiated, while reflective abstraction translates them explicitly into relations, *d* and *t*. Hence, it is by composition, that these various aspects come to be represented in the equation $v = d/t$. This composition is based on the correspondences expressed in our illustrations I and II. They can therefore lead to new observations within which the arrival points come to be incorporated. To be sure, a whole new process of generalization is necessary in going from these comparisons between completely or partially synchronous motions, where empirical observations are sufficient, to those between successive motions, which require computation and metric quantification. But, here we only discuss the more elementary levels.

2. Let us now consider the relations between linear and angular velocities. For example, one might ask what the relation is between the distance covered by a wheel and its speed of rotation or between distance and the number of rotations. The latter relation concerns speed and frequency. As for the former, it is only at about seven or eight years that the child begins to understand that one turn of a wheel covers a fixed length of distance, which is independent of its angular velocity. Before then, children think that if a wheel covers ten centimers in turning fast, it would cover only five if it turns slowly. In addition, if a wheel makes two turns in a row, it goes "faster" than a similar wheel which only turns once, even if the former turns more "gently": in this case, speed is seen to be a function of distance covered only, regardless of time and angular velocity. At about seven or eight years, however, subjects discover that to one turn of the wheel there corresponds a fixed linear distance, the speed of rotations being of "no importance" in this respect (subject Ari at 9;2 years). They also note the inverse relation between speed and duration. But not until eleven or twelve years are the various relations arrived at by deduction, at which point they are thematized: thus, linear velocity becomes "the distance covered during a certain time" (Syl at 11;2 years), and angular velocity "the time taken by the wheel to make one turn" (Ant at 12;0 years). Thus, it seems clear that these ultimate generalizations are not the result of simple additions of empirical abstractions and that the composition of the relationships involved requires that reflective abstraction play an active role.

3. This is even more true in the case of the velocity–frequency

relation considered alone. Our experiment was as follows: through a narrow vertical opening in a screen behind which a disk rotates, the subject sees a red line appear at various intervals.

At an initial level (four or five years), the subject evaluates speed only as a function of her own perceptions: "it goes slowly" when one can see it well or "that went real fast," because "one cannot see the red line very well"; in other words, the mobile is faster than the eye movements. At Level IB (six years), subjects still think in terms of being able to "see it well," sometimes including a temporal reference (have a "longer look"), but still making reference to their own actions. Frequency begins to be considered, but only in spatial terms: "you could see red almost everywhere" (subject Jac at 6;10 years).

At Level IIA (seven to nine years, the age when concrete operations form), global judgments explicitly refer to frequency: there is greater speed when one sees the red line many times. But the notion of duration remains implicit (more or less "often"), and when one tries to have the subject specify this notion, she refers only to a single phenomenon: the red returns "faster" or "more slowly" than expected. Even at Level IIB (nine to ten years), when we ask subjects to count the number of times the red line passes or to measure the time (with a stopwatch), they either take this measure without paying attention to frequency (even concluding frequently that "faster = more time") or else they count without taking the time.

However, they also often succeed in finding an adequate method, which is to measure the time it takes for the red line to reappear: in this case, two seconds means, according to the subject Met (10;4) that "it turns faster" than another rotation which returns every six seconds. But it can be seen that here frequency is relegated to a secondary role, at the expense, as at the preceding level, of isolated appearances as opposed to their sum.

At Level III (eleven to twelve years), frequency and durations are finally coordinated explicitly: "You turn for 15 seconds and count the marks," says Man (12;4); speed is then "the number of turns within a certain time interval." Here again the final generalization appears as the product of reflective abstractions directed at the operational instruments which had previously made the empirical abstractions possible in so far as these had been correct.

4. Let us consider a last example in which two kinds of independent velocities are involved.[7] This is a situation in which two cogwheels of different sizes are interlocked. The smaller wheel performs two and a half turns, while the larger one turns only once, while the perimetric paths, of course, have the same speed for a given sector since there is interlocking. The child has the option of using a series of cues to help her note the facts.

The preoperational subjects see no problem (stage I, from five to seven years), since they pay attention neither to the number of turns nor the number of cogs: they content themselves with perceptual evaluations, which remain subjective. During the entire stage II (seven to eleven years), however, subjects do distinguish two velocities, but vacillate between the two or resort to contradictory compromise solutions: "the small one gains more time," but the speeds are the same "because they turn together." "The big one makes fewer turns, it turns more slowly, but they both go at the same speed" (Mag 9;4). "The big one turns more slowly, but at the same speed as the little one. The small one turns faster, but it has the same speed: it depends on the size of the wheel, because it is smaller and has fewer cogs" (Mia, 11;7). Finally, at stage III, the velocities are differentiated and coordinated: "The small one turns faster about its axis, but at the same speed at the cogs."

"So, there are two kinds of speed?"

"Yes, faster around its axis, but after all, it cannot jump teeth" (Gad, 12;5).

Thus, it is clear that in this situation of interlocking gears, the subject has to solve two sorts of problems: (1) To differentiate the two types of velocities involved, one being angular and concerning the number of turns of each wheel, the other relative to the number of cogs within a common sector of perimeters; (2) To integrate these velocities within a general form, and from this establish the relationship between certain displacements and the time required. Now, while the reactions of Level II illustrate the part played by empirical abstractions in the emerging differentiation, they also show in what respect the process is still insufficient, since without a coordinating reflection the distinctions lead to contradictions: that is, the two problems, differentiation and integration, have to be solved simultaneously in this case. This is accomplished by a synthesizing generalization and reflective abstraction directed toward the composition of the relations employed previously.

5. In conclusion, the general idea of velocity developed by stage III subjects (eleven to twelve years) results from their establishment of relations between a varied and heterogeneous set of changes and a common denominator. These changes presented themselves as (a) linear displacements (b) angular displacements, (c) successions in linear order (the number of turns performed by wheels advancing on a table surface), (d) successions without displacements (frequency of lines appearing at a slot as in section 3, and (e) meshing gears. Subjects discover that in each of these instances duration[8] is important. Thus, as one of our subjects remarked, "there is always time" would seem to be self-evident, and in a way it is surprising that so much trial-and-error is necessary to find this solution. The question arises then what this duration consists of. For subjects from a certain age on, it can easily be read from a clock, but this does not take care of speed, since one needs to set up correspondences between the preceding speeds and the hands of the clock. Thus, there seems to be a circle, and if there were no clocks, this circle would indeed be vicious: duration, implicit or explicit depending on the situation, represents, in all cases and at each level, events with a certain content, but only with respect to the speed with which they occur. In other words, if $v = d$ (or $n)/t$, we have reciprocally $t = d$ (or $n)/v$; we have given ample evidence that this relation can be found at the psychological level. Therefore, the various forms of change noted by our subjects give rise to evaluations of speed and of time simultaneously. This is undoubtedly the reason why it is so difficult and takes so much effort to interrelate the two. In fact, while it is true that there is a circle between speed and time, it is not a vicious one; it is based on a symmetry: that of the correspondences expressed in our schematic illustrations I and II (in section 1). While subjects come to correct estimations fairly rapidly, in many cases in basing their judgments only on temporal and spa-

tial order relations (or on frequencies of succession), it remains for reflective abstraction to bring about thematization, operating on interval relations and coordinating the two symmetrical systems (which so far have been considered separately or in alternation), of space or frequency and of duration. Thus, we find again the usual combination of empirical and reflective abstractions.

But, in this particular case, there is more: if it is true that the relation $v = d$ (or $n)/t$ is general, its terms must be newly constructed for each domain or subdomain considered. Furthermore, in case (4), two independent types of speed are involved. This may appear contradictory, except where the two cogwheels have the same diameter, in which case the two domains would intersect. Thus, we are dealing here with an object with permanent properties, while velocity and duration are variables that have to be determined for each new context.

3. Constructive Generalizations[9]

A third example is useful to show the way physical generalizations, from their most elementary forms on, tend to transcend their initially inductive and simply extensional forms (from "some" to "all") in the direction of constructive and completive forms. This happens in so far as logico-mathematical structures are attributed to the objects themselves and are considered as operators. This transformation of generalization thus amounts to a substitution of the extrinsic variations of phenomena, observed through exogenous pathways by systems of intrinsic variations. These are logically necessary or at least necessary openings towards particular possibilities, which are inferred or deductively constructed through endogenous paths. A simple but eloquent example of this process in physics is the evolution of crystallography: on the basis of empirical observations of the geometrical forms of crystals, groups of transformations were constructed from which these forms could have been produced. As a first approximation, thirty-two have been identified, which are all realized in nature. Thus, a system of intrinsic variations came to replace and to subsume entirely all the extrinsic variations initially observed.

An example of the change from extrinsic to intrinsic variations at the elementary level is provided by children's reactions to the mobile effects of superpositions based on the principle of "moirés" or wavy effect. A card *F* serves as a stable base. It contains parallel lines which may be presented vertically, horizontally, or slanted right or left. A transparency, *T*, containing the same lines, is then placed on the card. However, the lines on the transparency are presented only vertically or horizontally. Various effects can be obtained by moving the transparency over the card, with the motion coming to a stop in various positions or applied in a particular position, as follows: (1) The lines on *T* coincide with those on *F* and the figure remains unchanged; (2) a juxtaposition, where the lines of *T* appear in the empty spaces of *F*, hence a dark figure; (3) if the lines on *F* and *T* are both vertical, a continuous movement of *T* over *F* produces alternations of black and white; (4) if the lines of *T* are perpendicular to those on *F*, one obtains squares; (5) if the lines on *F* are oblique and those of *T* are horizontal, one obtains stationary diamonds oriented one way or another; (6) if, on the other hand, the lines on *T* are vertical while those on *F* are oblique, moving *T* over *F* produces apparent movements, a kind of a waveform with displacement of the diamond upward or downward and deviations to the left or to the right.

With this material, one observes continuous development beginning with the simple reading of effects, which are neither anticipated nor understood, consisting only of extrinsic variations toward gradual deduction and composition such that each stage opens up new possibilities for the next levels up to a closed system consisting of the intrinsic variations we described before. The point of departure is completely empirical or "extrinsic" so that even children of four or five, having observed superposition (1), where the horizontal lines coincide with those on *F* going from left to right, begin to suspect that the effect would be the same in the opposite orientation: "Perhaps it would be the same, but I don't know," said Mag at 5;2 years. But, we do not find any predictions of the facts of alternation, as in (3) or even of juxtaposition (2). However, once these variations are noted, subjects of about five or six come to predict squares. They do this by means of a figurative imagination of the results obtained when one grid is superimposed on the other with the lines oriented perpendicular to each other.

When they are given two or four small rulers with the instruction to reproduce in detail the patterns observed or even predicted, the subjects falter a great deal more. In other words, they do not yet make use of deduction based on the movements but simply carry out figural superpositions of two sets of lines. In contrast, at the operational level, seven to eight years, the possible variations of the figure begin to become intrinsic, in as much as they are based on combinations of movements that are computed or deduced, accompanied by adequate explanations, in the case of two or four elements. But in spite of this great progress, it is very instructive to note how slowly generalization develops, advancing only step by step, focusing at particular questions without seeing the total array of possibilities: in other words, isolating each question without attempting to find a structure. Thus, at Level IIA, the predictions depend upon the various positions of the lines on *F*, and subjects do not take into account that there are two orientations possible on *T*, according to the direction in which *T* is moved, either in the same direction as the lines on *F* (that is, horizontally) or in a direction perpendicular to them (that is, vertically). Similarly, they do not yet succeed in decomposing, within a diamond or even a square, the sides that come from *T* and those that belong to *F*.

Only at Level IIB (nine to ten years) do subjects come to dissociate the different variations and to deduce them. Yet, this result is obtained only by trial-and-error methods. In particular, subjects are now able to show, by means of four squares, that the diamonds "like that (vertical *T*) they go up, and like that (horizontal *T*) they do not move" (Rol 9;6). Still, these apparent movements are considered real in the sense that the diamonds are considered as permanent objects each of which conserves its four sides when it gets displaced. At Level III, finally (eleven to twelve years), all possibilities are anticipated and explained and the apparent movement is characterized as an "impression": "The lines of *T* will go up . . . all at the same time over the lines of *F*: one has the impression that there is a crossover that rises up."

"Are they are always new ones or are they the same ones?"

"They are new ones, because it's not always the same lines that pass" (Per, 11;5).

Another case: "The angle disappears more and more (gets displaced), so it goes up"; and "the next line arrives where the others were before."

This development, where, finally, apparent movement gets relativized, is noteworthy for several reasons. First, it is clear that the variations of the device begin with an "extrinsic" form; that is, they are simply empirical registrations. There is no anticipation or explanation after the event. At the end of this evolution, however, we definitely find intrinsic variations, i.e. variations that are logically implied by the relations of the system and deduced by the subject in exhaustive fashion, thus covering the entire set of possibilities compatible with the relations. The second point of interest in the behaviors characteristic of the levels described is that these possibilities are far from being perceived all at once in one block. Rather they are the result of new possibilities opening up successively, very similar in kind to the variations already discovered. The perception of new possibilities is not immediate; there is a negative or if one wishes, a dialectical dimension involved. There is a stepwise elimination of preceding limitations—the negation of earlier exclusive and unique possibilities in favor of other slightly different ones. We have here, then, a composition of affirmative with negative statements that can be expressed in the words "not only x but also x'." Once subjects understand that superposition on F (effect 1) is the result of moving T over F in a horizontal direction, they also begin to realize that this movement does not only generate this particular effect but can also produce juxtapositions. If T moves across F, this movement can also occur in a direction perpendicular to the lines, and not only in the same direction that coincides with the lines of F. This is also true for other orientations, such as oblique ones.

From the point of view of the history of science, it seems interesting to compare the length of time it takes subjects to discover a possibility on the basis of a preceding one and the often considerable intervals of time separating the discoveries of a precursor from final achievements. Even if one believes in Bachelard's "epistemological cuts," it still remains problematic why a scientist who discovers certain partial relationships would not immediately see at least some of the possibilities they render available, even on a very limited scale. Why is it, for example, that Apollonius used tangents and their corresponding diameters as coordinates in his study of conic sections, but did not go on to construct other interfigural relations—even without having algebra at his disposal? Or, to stay with a scientist whose work maximally represents all the symptoms of a "discontinuity," why did Darwin, as Gruber[10] shows, take so much time to develop his ideas in the direction implied by his earlier ones? There seem to be two reason for this. One reason is that in order to solve a particular problem, one concentrates on the data so that empirical abstractions (or "pseudo-empirical" in the case of data concerning operations) and inductive, extensional, (hence limited) generalizations predominate. To achieve constructive generalizations and to overcome these limitations ("not only . . . but also"), however, a complete change in direction is needed and the use of reflective abstractions is required. This permits the discovery of the operations previously used instrumentally as objects of explicit thought. Such changes in direction or reversals are not easy, and they never happen all at once.

A second reason for the length of time between discoveries and final achievements is that a set of possible intrinsic variations constitutes a structure, and it is well known that there is always a delay between the use of operations and their constitution as structures. The reason is that structures require a higher level of reflective abstractions and completive generalizations because they require compositions between operations: what is involved here are operations on "components," and this requires a new equilibrium between differentiations and integrations, including closure.

In the particular case of our fluid patterns described above, such closure is achieved when subjects come to understand the reason for the differences between the situations producing static figures and those producing apparent movement, which are then explained in terms of a relativization of movements. In fact, such relativities characterize all physical structures in which several systems of reference are operative. What is interesting about these is that they permit the distinction of two kinds of constructive generalizations, where the second is hierarchically superior. The first amounts to the integration of a system, already known, as a subsystem within a larger system; however, there is no retroactive enrichment of the earlier system. For example, after having understood the way squares are generated, subjects generalize this to the case of diamonds in such a way that squares are included as a subsystem within a new, more extended system. But the preceding system remains unmodified. In contrast, subjects who understand the relativity of the apparent movements do not simply generalize to these cases what they learned from the earlier situations. Rather, they enrich these retroactively by discovering that all the figures produced, whether static or kinetic, can be explained by a single general principle.

4. The Reinterpretation of Variables

After having examined the alternation of empirical and reflective abstractions and their relations to the two types of generalization, both in identical and different domains, it remains for us to find the similarities, much less obvious, between the history of theories and the psychological emergence of concepts in situations where variables have to be reinterpreted and subordinated to identical principles of conservation, extended to new domains.

Since Newton's times, it has been possible, in physics, to distinguish two kinds of processes with respect to this phenomenon. The first is characterized by the fact that a single principle of conservation, at first thought to be sufficient, then gets completed by a second, more or less analogous one, where both are experienced as necessary, either jointly or alternately. For example, it is generally known that the Cartesians explained the transmission of movement in terms of the conservation of $mv/2$ only, while the adherents of Leibniz's theory did so in terms of the invariant mv only (which became $1/2mv^2$); both of these are different but necessary consequences of Newton's law.[11] The second process comes into play when a single conserving expression such as Hamilton's function H (which itself resulted from a reinterpretation of variables from Newtonian and Lagrangian physics concerning velocity and quantity of motion) gives rise to reorganization of variables in order to be applied to other domains, such as microphysics. Now—however bold this comparion may appear, and however far removed children's reactions may be, when first being confronted with such problems, from the slow process by which physical theories were elaborated—it turns out that the sequence of solutions observed at different developmental stages resembles *mutatis mutandis* the two processes just described.

1. First, with respect to a single principle of conservation which then becomes divided in two, we can cite certain behaviors that are interesting from the point of view of "natural" logic (taken in the sense of "natural" numbers with reference to a their prescientific genesis). When subjects of eleven or twelve—an age where conservation becomes generalized to all domains—are confronted with easily decomposable transformations of surfaces and perimeters,[12] they tend to affirm that both are conserved, as if the visible modifications of the two dimensions should necessarily compensate from both points of view at once. Yet, one of the figures in our experiment had for a perimeter a nonelastic thread in the form of a loop, and the child varied the angles by means of pins. In this case, the surface area decreases so obviously with the increase in length of the long side of a rectangle that it gets reduced to zero when the two sides of the thread make contact to form one double line. Nevertheless, even subjects past the age of eleven declare (as does Cha at 11;9 years) that, in this limiting case, the surface remains the same and that it is located "between the two lines"—that is, in a space that has become invisible. Similarly, consider a case where a rectangle is formed out of eight cardboard strips, which are first adjusted to form a square, then arranged 4 × 2, etc. until one obtains an alignment 1 × 8. The perimeter, increasing by a factor of four from the initial situation, is now considered to remain constant by compensation between length and width: "The sides, the surface, that has not changed," says Geo at 11;2 years.

"And the perimeter?"

"There is more in length and less in width so that come to the same" (Cal at 11;5).

What we are dealing with here is a global form of conservation, based on an inference that is correct in principle, but generalized in total disregard of the most obvious perceptual data. Progress toward a solution then consists in dissociating the variables, which were distinct until they were falsely linked together out of concern for logical unification, and then applying to each separately the deductive schema of compensations. This is, for example, what one eleven-year-old does in concluding: "When the surface changes, the perimeter does not change (the first figure), and when the surface does not change, the perimeter does (second figure): they both change, but not at the same time."

We have here the example of a single structure of conservation, applied to two classes of variables after they are reinterpretated as varying independently of each other. We have cited only the responses of the upper levels (nine through twelve years), where reflective abstraction outweighs empirical abstraction, which gets resorted to only for the sake of checking out facts. At lower age levels, when empirical reading of fact still dominates deduction, the answers, curiously, are better as far as their content goes. But they are less interesting: they do not yet show this concern for systematization of all compensations simultaneously, which is what leads the subjects of the next-to-last level into error.

2. The following facts are relative to a case where a concept of conservation is enlarged by the reorganization of variables to be added under a new form to those used previously. This is a complex device which is far from being immediately comprehended by the average adult (or even by a college student not specializing in physics). Two pendulums[13] are suspended in parallel. The wires by which they are suspended are connected horizontally by means of an elastic band: when the first pendulum *(A)* is swung its movement gradually gets transmitted to the second *(B)*, but as the action of *B* increases, that of *A* decreases; what happens—and this is the main problem—is a reversal of roles such that the second pendulum, now active, activates the first pendulum, now passive, into motion. This alternation continues repeatedly between *A* and *B*; hence there is an interchange of energy transmission (which could still be explained in terms of pushing and pulling).

During a first stage, subjects succeed, after they have noted the facts, in understanding that movement was transmitted through the elastic, but they neither predict this transmission nor do they see therein a form of conservation, since they explain the reactivation of *A* simply with reference to the fact that, following the oscillation of *B*, *A* "regained its momentum," as if the impetus could simply get lost and reconstituted, etc. During stage II, subjects predict the transmission and explain the reactivation of *A* once it is noted—but not yet predicted—in terms of reciprocity. At stage IIB (ten or eleven years), this reactivation begins to be predicted, but not its continuation. at stage III at last, some subjects (from twelve to fifteen years) predict and explain the entire process, which is remarkable.

Now, we have here in fact two distinct kinds of transmission, giving evidence of an extension of conservation: a simple transmission of movement from *A* to *B*, and second, an alternation of the active and passive roles of elements *A* and *B*, which amounts to the transmission of a "capacity," and not simply of motion. What happens—and this is striking on the basis of predictions and before subjects are able to make any kind of observation—is a reorganization of observables leading to the idea of an alternating delegation of capacity, which is conserved while changing carriers. Without necessarily wishing to attribute to the subjects a very clear notion of energy, the fact remains that this is an example where an initial form of conservation is extended, requiring a new way to manipulate variables and a dissociation of two kinds of invariants.

III. THE VERIFICATION OF TRANSITIONAL PHASES IN THE PSYCHOLOGICAL DEVELOPMENT OF THE NOTION OF WEIGHT[14]

A. Weight is a notion whose development is quite complex since, as a property of objects (nondifferentiated from their mass), it does not constitute a simple observable, but depends both on the quantity of matter and on (even more than on its density) the presence or absence of "empty" spaces within the object. A further complication is that it can be the source of multiple actions, of which the fall of bodies is by no means among the first ones to be discovered. In addition, it constitutes a good example of the process of the intra, inter, and trans phases with respect to the interpretations subjects successively construct. We shall designate these interpretations *T1*, *T2*, etc., as if they were theories. Let us first give a description and then attempt to find the formative mechanism.

The first interpretation, *T1*, is one in which observables or their interrelations are merely described: weight is simply a function of size, which is evaluated in an ordinal way, without operational additivity. It is thus seen to vary with position (and without conservation between two weights presented in piles or juxtaposed, suspended by means of long or short threads, etc.). It is also seen as varying with speed in its various effects (intensity of shock or resistance, etc.). It is not understood as exerting a fixed amount of pressure (as we indicated above in section II-B): The equilibrium of a scale with two arms of equal length is seen as due not to equal weights, but to their symmetry. Subjects believe that this equilibrium would be lost if two equal weights of differernt shapes were used, or even if one placed two blocks on one arm and two similar ones on the other instead of one against one.

A model *T2* appears at about seven or eight years, with weight becoming additive—that is, conservation under positional changes without any changes in shape. As a result, in simple cases (that is, before children understand the notion of "momentum"), the equilibrium of a balance is interpreted as being due to two equal weights acting as two forces in opposition. But the weight of an object is not conserved where there is a change in shape (e.g., a clay ball getting transformed into a sausage). Hence, weight is not understood to be proportional to the quantity of matter (which does get conserved). (All these notions are acquired with *T3*). Most importantly, weight is not yet associated with vertical fall, as can be seen in residual confusions between downward pull and retaining: for example, subjects who operate with *T2* know very well that a board extending beyond the edge of a table can be held in place if a counterweight is put on the end resting on the table; but it sometimes happens that children add another weight, which they put on the protruding end in order to keep everything in place. Similarly, when one holds a stick so that it has an oblique orientation, some subjects predict that it will fall in the same direction rather than vertically. A train car resting on an incline of between 45 and 90 degrees requires more power to be held in place than to be pulled uphill, since it tends to roll down when stationary, while this tendency disappears when the wagon gets pulled upward.

All these different problems are solved with *T3*, where the conservation of weight is acquired in situations where there are changes in form, and where subjects understand vertical fall, constructing systems of natural coordinates. The most characteristic example here is the prediction and explanation of water levels. Before this stage, subjects believe that water is "light," being mobile (whereas ice is considered heavier, being solid), so they do not attribute the downward flow of a brook to weight. Rather they attribute the flow to the absence of obstacles on the downward side and their presence on the upward side. At about nine years, however, subjects see the descent of water as being due to its weight. They also predict that, when stationary, water will have a horizontal surface and that as soon as there is an incline the parts located higher up would descend on the lower ones.

But these advances in relating weight to vertical fall are bought at the expense of an apparent regression toward nonadditivity in the sense that it is believed an object is heavier near the bottom than near the top of a suspension, etc.; but, in fact, such hypotheses are due to an improvement in dynamics rather than to a failure to coordinate weight with quantity of matter. In fact, such coordination continues to be operative and in case of a conflict (especially when one raises the question of invariance), subjects resort to various kinds of compromise solutions: the weight remains the same, but it "gives," "pushes," or "weighs" more or less depending on the situation.

Proof for this assertion that the dynamic complications do not interfere with the additive property of weight characteristic of *T3* interpretations comes from the following observation: given four or more equal weights ($A = B = C = D$), subjects of a lower sublevel understand these equalities ($A + B = C + D$) but only for homogeneous objects (four similar blocks), whereas in phase *T3* (nine to ten years) they generalize this to heterogeneous objects (three bars of brass and one piece of lead). It is also essential to note that subjects at this phase, *T3*, succeed in performing, in operational fashion, seriations of weights $A << B << C$ . . . including transitivity (A << C if $A << B$ and $B << C$), while transitivity of lengths is already present at 7 years. Furthermore, we find composition of equivalences ($A = C$ if $A = B$ and $B = C$) between weights of heterogeneous objects. (A and B = two bars and

C = a piece of lead). This clearly shows the internal coherence of the general sense of the interpretations *T3*.[15]

But it is only at level *T4* (eleven to 12 years) that more general quantifications are performed successfully. This is possible because not only do subjects establish relations between different weights or the direction of their effects, but in addition they compose weight with different spatial dimensions of objects: e.g., volume for probems involving density and floating of objects, surface for problems involving pressure, and length for those involving momentum.

With respect to density, the subjects of level *T1* predict that weight varies directly with volume, while at level *T2* one already obtains responses like the following: "there are some big things that are lighter than little things" (Daf, 7 years), but only as a result of the intrinsic quality of the objects: "because this is cork and that is rock" (the same subject). As a result, children know that to make a clay ball of the same weight as a bottle top it has to be smaller, whereas at level *T1* they make them the same size or even bigger. However, to make a ball of half the weight of the bottle top, it is only at phase *T3* (nine to ten years) that subjects are able to divide their clay in two. Also at this level, unequal densities are interpreted as resulting from the more or less "filled" nature of the objects. Finally, at level *T4*, this notion is given up in favor of another concept of definitely higher rank: that of "crowding" measured on a scale, that is, now at the level of corpuscles. This explains the inverse relation between weight and volume which characterizes the increases in density. Let us further observe that at this level the expansion of a grain of corn slightly heated is no longer attributed to an increase in substance nor to simple "swelling," but to decompression: the weight remains the same, the number and the size of the "small grains" likewise, but, while before they were "all crowded together," the warm air drives them apart" (Jac, 12 years) and the volume increases.

With respect to floating objects,[16] the subjects of level *T1* limit themselves to descriptive accounts or invoke weight as a factor, but with contradictions due to their belief that weight and volume are directly proportional or to an ambiguity in the dynamic effects invoked. For example, subjects will say that small boats can float because they are light and the water can carry them while large boats can float because they are heavy and can support themselves. At levels *T2* and *T3*, weight becomes relative in two ways. The first is in terms of qualitative density: wood floats, because it is light and the rock sinks, because it is heavy, etc. The second is that subjects begin to see a relation to the volume of the water: "a big boat, said one child, is heavy for us but light for the lake." Finally at level *T4*, a body floats "because it is lighter than water, volume being equal: a key does not float, because the same contents of water would be less heavy than the key, said Ala" (11;9), while Jim observed that "It would take much more water than metal to make the same weight." Lamb even goes so far as to say "You mark the water level on the wall of the container (before immersing the object, a wooden ball), then you put the ball into the water and let out the amount of water that makes the difference with what it was before."

"What are you comparing?"

"The weight of the water taken out of the container and the weight of the ball."

This, in fact, is the principle of Archimedes, which, however, this subject had not yet studied in school.

As for pressure and its various relationships with surfaces, its relation to weight is also understood only at level *T4* at about 11–12 years. For example, for Luc (10;10), a bar of metal exerts more pressure "vertically, because the whole weight is there (at the square base), whereas the weight is distributed when it is horizontal." The reason is that this relation involves two conditions. The first is that the objects which exert or undergo pressure, must be considered as continuous (i.e., their different parts remain associated) so that the weight in its entirety is either "distributed," if the surface of contact between the objects is large, or "concentrated," if it is small (Yva 12;6). The second is the conservation of weight in the face of variations in the surface area to which it is applied. Now, before they reach this final level, subjects believe that an object extending beyond the edge of a platform on a scale exerts less weight on the scale than if the object is placed so that it does not protrude (we even encountered this belief among some adult grocers).

It remains for us to report the following observation: given a cylindrical container with a hole in its lateral wall so that water escapes from it, subjects up to level *T3* believe the water pressure that makes the water escape depends not only on the upper layers but just as much on those located below the opening.

As for the relations $P/P' = L/L'$ of the weights $P >> P'$ and the lengths $L << L'$ on the arms of a scale, it is also at a "transponderal" level that these are established and explained: "The greater the distance (from the center) the less weight should there be. . . . Distance and weight—that makes a system of compensation (Chal 13;6), and the reason is attributed to the work involved: "It takes more strength to lift up [the distant weight] than when it is closer to the middle" (Sam 18;8).

B. If we have thus recapitulated all these well-known facts, it is not in order to limit ourselves to showing that the sequence of interpretations from *T*1 to *T*4 provides an excellent, global illustration of the transitions from the initial intrafactual (*T*1) to the final transfactual (*T*4) through two progressive interfactual stages (*T*2 and *T*3): this is obvious and not at all surprising! The most interesting epistemological problem is this: since we are dealing here with physical facts, hence causal explanations rather than simply logico-mathematical structures, what transitional mechanisms can account for the progressions from a stage Tn to the next $Tn + 1$ in order to discriminate the relative contributions made by endogenous structures and exogenous data? In addition, is their interaction comparable to that what the historical-critical analysis (above) has shown? To orthodox positivism, knowledge of physical facts is nothing but an accumulation of exogenous information translated into a logico-mathematical language, which serves only to describe the world by means of a precise symbolism. In psychological development, when this symbolism is not yet acquired, the role of exogenous information should then be proportionately even more important. For this reason, our genetic-historical comparisons are of particular interest in this domain, where one might expect them to provide an answer to one of the most central problems of epistemology.

1. The first problem concerning the nature of interpretation *T*1, which we present here as the initial one (we cannot question infants about weight). To read our description, it seems as if this

pre-system *T*1 consists in nothing but a collection of observables that are momentary (except for the fairly constant relation between weight and size, which, however, is often denied), local (depending on the position of the object, etc.), and, to a large extent, inexact or simply erroneous. It may also seem as if these deficiencies are due to the complete absence of structuring activities on the part of the subject, since she does not yet apply any form of conservation, additivity etc., but, at best, certain perceptual symmetries: this would be tantamount to saying that the observables involved at this initial level are nothing more than the products of defective readings, persisting in a stage of pure exogenous recordings of facts.

Yet, this would be completely false, since, as we said in our Introduction, the "readings" are the product of assimilations to the subject's schemata. If there were only "copies" of reality in the sense of Hull's empiricism, this would raise the problem of how to account for cases where these readings deform reality. The characteristics of the interpretations *T*1 must then be sought in the interaction between objects and subjects' actions, where these actions are not yet coordinated within coherent operations, but limited to accept the effects they provoke or discover (in fact, subjects' manual evaluation of weights, etc. varies with the situation). Our interest in these pre-systems does not lie in their flaws (even though it is always instructive to note the initial errors to evaluate the difficulties to be overcome and get an idea of the long road ahead in the conquest of truth); rather, it is in the complex dual nature of the process, which relates subject and object (which is obvious), and also observables to inferential coordinations (this general term will have to be differentiated into abstraction, generalization and composition). For it is this dual circularity which accounts for the subsequent loops and alternations between successively reciprocal actions which characterize the usually nonlinear transitions between *T*1, *T*2, *T*3, etc., transitions common to the history of theories and the most elementary stages in psychological development.

In fact, if we call *Obs, O(n)* the first observables obtained from the objects as a function of the subject's action and *Obs. S(n)* what the subject learns about his actions, *Coord. S(n)* the way she coordinates these actions and *Coord. O(n)* the coordination between the *Obs. O(n)* the dual circle can be graphed as follows:

$$\begin{array}{c} O \rightarrow S \\ [Obs.S(n) \rightarrow Coord.S(n)] \rightleftarrows [Obs.O(n) \leftarrow Coord.O(n)] \\ S \rightarrow O \end{array}$$

This means that the subject becomes conscious of her actions only through the results they produce on objects, but understands objects only by means of inferences derived from coordinations of the same actions. Now, *Obs.O(n)* and *Coord.O(n)*, when combined will generate new *Obs.O(n* + 1) and *Obs.S(n)* + *Coord. S(n)* new *Obs.S(n* + 1), *Coord.S(n* + 1), and the cycle starts over again.[18] It is this complex situation which explains the formation of *T*1. It will be sufficient, in accounting for the change from *T*1 to *T*2, etc., to make the following distinctions, which correspond to the differentiations and constructions carried out by the subject herself:

a. The observables *Obs* will be of two kinds, depending on whether they are discovered by empirical abstractions or constructed on the basis of coordinations, by reflective abstractions, subject to verification by new experiences.

b. Coordinations also are of two kinds depending on whether they proceed on the basis of constructive generalisations or by extension of the domains *D* of application.

c. The interpretations *Tn* can then be interrelated by means of operational compositions, which equilibrate differentiations and integrations.

In brief, this is the way generalized schemata pass from one stage to the next in the evolution of historical theories, as sketched above.

2. To come back to the psychogenesis of the notion of weight, we have to find a way to account for acquisitions as well as for deficiencies characteristic of interpretation *T*2. The former are easy to understand, since what is involved is the notion of additivity of equivalent weights and their conservation. Both are independent of the relative positions of the weights (superpositions or juxtapositions, etc.), as long as the objects involved do not, individ-

ually, undergo changes in form. Two preliminary remarks have to be made in this regard. The first is that if one has, for example, four blocks $A = B = C = D$, it takes only simple experience (with, say, a scale) to show that $A + B = C + D = A + C =$ *etc.* Now, the child does not start from there. If he does, it can be observed that he deduces the results before controlling them. Hence, a second essential remark: in order to have such experiences, it is necessary to conceive of them and to be able to program them. Now, such advances, inaccessible at Level *T*1, presuppose that the concepts of conservation and additivity have previously been minimally mastered, in terms of new possibilities. The problem is then to understand the process that makes them available as possibilities, to prove that this is indeed a new process, and to explain how it is that precisely at the time when the subject succeeds in conceiving such possibilities, he immediately carries them out in thought so that their outcome is postulated as being logically necessary, no experience being needed.

It is, thus, easy to see the important role played by reflective abstraction. To begin to conceive of objects (*A*, *B*, *C*, *D*) that are apparently equal in form and quantitative values as constantly changing their weight, depending on situations and positions, is to accept a complicated system of continual modifications, where gains and losses can be explained locally but are difficult to coordinate: *A* on *B* weighs more than *A* side by side with *B* because of the additive effects of pressure, but, if *B* on *A* is equal to *A* on *B*, why deny their equality in the situation of lateral contact, where both weigh on the same support? In short, as soon as subjects are able to explain modifications separately, they will try to coordinate, attempting to solve the two problems inherent in any system of transformations: (1) how to separate the elements that vary from those that remain constant; and (2) whether or not compensation is involved in these variations. Thus, it is clear that without any measurement or computation, the simplest hypothesis is that for objects of equal shape and size, where only the position changes, the invariant is none other than weight itself. This makes it possible for the subject to engage in activities that are much more general and intelligible than the uncoordinated actions performed previously. This activity leads to additivity by means of composable and reversible operations. This is the nature of the concepts that are constructed between *T*1 and *T*2 (during the inter phase), leading to new empirical observables in *T*2.

As for the deficiencies of *T*2, while it is the case that weights have become additive, it remains to specify in what direction and precisely which positions on the support are affected by the weights. These are problems of a very different order. A weight exerting pressure downward can just as well retain as it can precipitate. We have already discussed the confusions that persist in *T*2 concerning the effects of counterweights: A weight that gets elevated no longer tends to descend the way it does when placed on a grade. Its descent is not always vertical, *etc.* In fact, these cases involve spatio-dynamic questions, which require new operational constructions for subjects to master them.

3. The transition from phase *T*2 to *T*3, which is characterized by conservation of weight in the face of changes in form of continuous stimuli and by the ability to handle problems concerning the direction of impact made by weights, is particularly interesting because everything that is acquired at phase *T*3 may appear to constitute the product of purely empirical abstractions or of simply relational generalizations on the basis of properties already known at phase *T*2.

To begin with the conservation of weight in situations of changes in form, it may appear that a subject who accepts that four blocks will always have the same total weight $A + B + C + D$ whatever their spatial arrangement, should immediately understand that a clay ball transformed into a suasage should conserve its weight, since it would be sufficient to section it into parts and to arrange these differently to be sure of the equivalence of their sum from the initial to the final situation. Now, it turns out that to get to this point subjects find themselves obliged to engage in complex constructions, within which reflective abstraction and operational compositions play a necessary role. First, they have to understand that a change in form can be reduced to displacements of parts, and this is far from being immediately accessible, since to do this one has to conceive what is continuous as being decomposable into "pieces" separable in thought, even though, in reality, there is only continuity without any visible frontiers. Second, one has to accept "commutability" (see chapter 4, Section III) inherent in displacements, in other words, the equivalence between what is first taken away and added in the end. Third, one has to generalize this process "vicariantly," so that it applies to any sections and displacements whatever. To introduce a formula already used elsewhere, we have then:

$$[(aA1)\, v(A'1) = (A'1)\, w(bA1)]$$
$$\leftrightarrow [(aA2)\, v(A'2 = (A'2)\, w(bA2)] \leftrightarrow etc.$$

where $(aA1)$ and $(aA2)$, *etc.* are mobile parts in their initial condition, where $(bA1)$ and $(bA2)$ are the same parts in their final states; where $A'1$ and $A'2$ are the sectors left in place during the transfers and where v and w are the unions of $A1$ with $A'1$ or of $A2$ with $A'2$, at different points.

It can thus be seen how this relation of commutability (expressed by the equal sign) and of vicariance (expressed by the arrow $\leftrightarrow$) implicates, to an important degree, reflective abstractions and endogenous compositions, although the final outcome—conservation as such—has become a new observable, accessible to empirical verification by means of scales. (However, this does not lead to any generalization, understood as logically necessary, as long as the subject remains at this empirical level.)

As for the child's mastery of spatial directions (water descending or remaining horizontal, as explained with reference to weight), one might similarly assume that these are simple empirical discoveries. In fact, none of these achievements are possible without the construction of coordinates and the endogenous compositions and particularly the reflective abstractions this construction requires.[19]

The role of the subject's operations in the structuring of the weight concept characteristic of level $T3$ can also be seen very clearly in the transition leading from $T2$ to $T3$—the inter transition—in the progressive constitution of weight seriation with transitivity, of generalization of additivity and transitive equivalences between the weights of heterogeneous objects. If only extensional inductions of an empirical nature were involved, it would be difficult to understand why these acquisitions appear so late in development, whereas the same operations are generally present at about seven to eight years in other domains. Given the hypothesis of an endogenous restructuring of a set of empirical phenomena, which are more difficult to master because of their dynamic properties, this developmental delay is significant.

4. With the different models of type *T*4, the advances finally achieved can be interpreted more systematically in a formula that covers all the facts noted: this is the coordination of weight with the spatial variables. In every case, this coordination is synthetic in nature. In contrast to the simple functions of *T*3, it leads—as we have noted—to new concepts of a general significance.

Among the meanings these new concepts carry, there is first the notion of density, understood as the constant relation between weight and volume. Further, there is the notion of pressure in relation to the surface, "momentum," arising from the composition of weight with length, as in the case of the two arms of a balance. We can even include the case of floating objects, where the explanations given come close to the principle of Archimedes. We shall not return to the facts described above; the conclusions we can draw from them are important here: with respect to the transitions discussed here, none of these concepts, which become new observables with the possibility given at Level *T*4 for empirical verification, could have been constructed without the contribution of endogenous constructions and particularly reflective abstractions. For example, density might appear to be given empirically from the initial stages on, when children say, about two objects of the same size, that iron is "heavy" and wood is "light." But to get from there to the notion of more or less "dense" ("squeezed") and to the corpuscular structure implied by it, there is the whole difference separating global intuition from a spatial structure furnishing "reasons."

In general, the contribution of endogenous constructions increases steadily between *T*1 and *T*4. It is marked by two kinds of effects which tend to become complementary. First, there is progressive structuring of reality: after the empirical abstractions of the "intrafactual" phase, which furnishes the initial data, there follow the inferential constructions based on reflective abstractions and completive generalizations, which establish relations between the data in the interfactual manner as well as with new notions, which are first deduced, but then reveal, in the following phase, the existence of new observables. This leads to an extension of the domain of knowable phenomena in the form of new "laws." But as this "lawful" structuring of reality brings increasingly into play the subject's logico-mathematical operations with an internal necessity of their own, there derives a second class of effects: the "trans-factual" relation established between a theory Tn and the next $Tn + 1$ becomes explanatory to the extent that the subject's operations are "attributed" to the objects, where this attribution consists in introducing necessity into the factual relations and not only into the implicative relations between a more general law on the one hand, and the more particular laws it contains on the other.

To the extent that this analysis is exact, it does not seem at all irrational to compare the transitional mechanisms in successive physical "theories" (interpretations), between the elementary levels of psychological development and the upper-level models and theories.

IX

Science, Psychogenesis, and Ideology

In the preceding chapters, we have tried to show in what respects the developmental mechanisms operating in psychological and development (regulating the child's intellectual development) resemble the sociogenetic evolution (relative to the evolution of leading ideas, conceptualizations, and theories) in certain domains of science. Until now, we have focused our attention on the problem of the knowing subject—that is, on the individual who assimilates the elements furnished by the outside world. In this process of assimilation, the subject selects, transforms, adapts, and incorporates these elements within her own cognitive structures. To do this, she also has to construct, adapt, reconstruct and transform her cognitive structures.

We have tried to describe this process by exposing the internal laws which regulate the dialectical interaction between objects that get incorporated into knowledge and the cognitive instruments which make this incorporation possible. But our study would be incomplete without a reconsideration of our analyses from a different point of view. Instead of focusing on the individual, we shall center our attention on the elements which constitute the objective reference of knowledge. In what follows we shall show that this change is necessary, in order to arrive at a total synthesis that can serve as an explanatory schema for the interpretation of the evolution of knowledge, both at the level of the individual and that of social evolution.

I. THE SOCIOLOGY OF SCIENCE AND THE SOCIOGENESIS OF KNOWLEDGE

In the Introduction, we indicated that the empiricist thesis is untenable, since there is no such thing as "pure" perception or experience. The "reading" of experience requires the *application* of cognitive instruments—which make it a reading—as well as the *attribution* of relationships between objects—which furnish the causal links between events. The intermediary between objects or events and cognitive instruments is action, as we have seen on many occasions. The manner in which action participates in the creation of knowledge, seen from the perspective of genetic epistemology, renders this position more precise, giving it its own identity with respect to classical dialectical philosophy, with which our position converges. This identity consists in our analysis of *praxis* into its constituent actions, which thus appear as essential factors in the initiation of the process of knowing.

Now, action does not take place only as a result of internal impulses (except for the first part of the sensorimotor period). It is not generated exclusively in centrifugal fashion. Rather, in the experience of the child, the situations she encounters are generated by her social environment, and the objects appear within contexts which give them their specific significance. The child does not assimilate "pure" objects defined by their physical parameters only. She assimilates the situations in which objects play a specific role. When the system of communication between the child and her social world becomes more complex and enriched, and particularly when language becomes the dominating means of communication, then what we might call direct experience of objects comes to be subordinated, in certain situations, to the system of interpretations attributed to it by the social environment. The problem for genetic epistemology here is how to explain in what way assimilation remains, in such cases, conditioned by a particular social system of meanings, and to what extent the interpretation of each particular experience depends on such meanings.

The history of science undoubtedly provides the clearest example of the influence of the framework of meanings in which society inserts objects and events. Unfortunately, the same problem with respect to the child's intellectual development must remain on more speculative ground, given the absence of experimental data.

1. Epistemic Framework and Paradigms

In chapters 1 and 3, where we analyzed the nature of the scientific revolution in the seventeenth century, we showed that the fundamental contribution of those who were responsible for this revolution was not in methodological refinement or in a substantial perfection of the instruments of observation, but rather in a reformulation of the problems constituting the object of scientific study. The revolution in mechanics was not the fruit of discovering new answers to the classical questions about motion, but resulted from the discovery of new questions, which permitted us to formulate the problems in a different manner.

It is from this point of view that we characterize the scientific revolution as a change in "epistemological framework." This concept of "epistemic framework" we are introducing is different from what Thomas Kuhn means by "paradigm." Let us briefly review Kuhn's approach[1] before we develop our own conceptions. We shall then discuss in more detail Kuhn's theory as well as the debates it has provoked. In this way, we shall be able to better situate the differences in perspective. Kuhn's theory of scientific revolutions specifies that each period is characterized by what he calls a "paradigm"—that is, a particular conception of what the ideal type of theory should be, the model to be followed in a scientific investigation. The criteria that determine whether a piece of research is scientifically acceptable, those which determine the lines of investigation (or, in the case of the universities, the thesis topics) which are likely to be approved, remain, according to Kuhn, in large part determined by the paradigm that is dominant at a particular time and place in history. Our notion of epistemic framework includes that of paradigm. Thus, our approach is not in opposition, but simply different. In fact, the concept of paradigm, as Kuhn understands it, belongs with the sociology of knowledge rather than with epistemology as such, while our concept is essentially epistemological.

2. Exogenous Factors: the Social Paradigm

The conceptual apparatus and the entire set of theories which form the accepted science at a particular time in history are the factors which determine predominantly the direction taken by scientific investigation. Certain trends of research take on importance, while others find little or no support. Certain topics become "faddish" and show hypertrophic growth at the expense of other subjects. All this usually takes place within a single epistemic framework, but may also bring about a change in this framework, when a subject gets elaborated to an extent that new discoveries are made either in instrumentation, making it possible to study problems hitherto inaccessible, or in the formulation of new questions, which modify the perspective in which investigations are carried out.

From this point of view, it seems clear that in the concentration of efforts in the investigation of certain phenomena, particular problems, involving particular research groups or certain types of observations, have a predominant role in determining the direction in which scientific theories develop. This concentration of efforts, in turn, depends on several factors. Stimulation and pressure come from social groups who demand solutions to practical problems. This is the case with technology applied to industry, the development of which has brought about discoveries which have opened new fields of scientific investigation. Defense technology is perhaps the most characteristic example. A good portion of mechanics—to cite a classical case—was developed under the impetus of the requirements of the artillery. A typical case is that of Euler's mechanics. Moreover, it seems reasonable to assume that nuclear physics would not have made one of its more spectacular advances without the stimulation and the powerful means allotted to it by the governments interested in using nuclear energy for military purposes.

It is evidently conceivable that if the stimuli had been different, other domains of science might have been the object of greater attention from the best brains of our times. Had this been the case, there would have been other discoveries, and different scientific theories would have developed to account for them. Thus, a large sector of scientific knowledge continues to grow, not in a strictly rational fashion in response to internal questions, but in a somewhat arbitrary manner, spurred by a set of impulses oriented by external requirements and imposed by society. It is for this reason that we shall call the type of paradigm that is conditioned in this way the "social paradigm."

It is certain that this process does not go only in one direction. It should not be forgotten that the basic ideas about nuclear energy were generated by pure scientific research, and that it was men of science—among them the most brilliant physicists of all history—who stimulated the military interest in this new form of energy.

The impetus for a spectacular progress in a particular branch of science can also come from a chance discovery or from advances made in other branches. Classical examples abound in the development of mathematics, where certain discoveries made it possible to treat certain problems in physics which had been inaccessible to theory. New themes and new fields of research may come to dominate the general interest and thus generate, among other things, new paradigms.

It is no longer necessary to dwell on these kinds of considerations, which have already been amply treated in the literature. We intend to use these cases as references in establishing some of the distinctions which we consider essential in the analysis of the mutual influence between science and society, in order to identify those elements which affect the development of cognitive systems, both at the psychogenetic and at the historical level.

3. Endogenous Factors: the Epistemic Paradigm

A first distinction we wish to establish, following the preceding considerations, is that between accepting or rejecting certain themes as being worth persuing and adopting or refusing certain conceptual schemata as being valid. The decision to invest a great deal of effort in nuclear energy rather than in the reconversion of solar energy is a decision in favor of certain themes of research, by virtue of their practical applications rather than for reasons related to a particular conception of an epistemic nature. That the preference for certain themes should have pushed science in a certain direction and that the choice of a different direction brings about developments that are likely to modify the whole outlook of contemporary physical theory are issues which seem to arise as a byproduct of the practical decision rather than from epistemic reasons. This is not to say that these questions may not have profound consequences for the development of scientific knowledge.

A rather different issue is that of the acceptance or rejection of ideas, concepts, or topics by denying that they are "scientific," at a given time in history, because they do not belong to the conceptual apparatus which the scientific community, by an explicit or tacit consensus, sanctions as the only valid one. Newton's mechanics had to wait more than thirty years before it was accepted in France. It was not criticized for any error in computation nor was there any experimental result that contradicted its theses conclusively. It simply was not accepted as being "physics," under the pretext that the theory did not provide *physical explanations* of the phenomena. It was the concept of physics as such that was contested. It is neither accidental nor insignificant that it was a philosopher—Voltaire—who played the main part in introducing Newton's ideas on the Continent. A few decades later, Newtonian explanations were not only universally accepted, but also became the definitive model of scientific explanation. In the nineteenth century, Helmholtz stated that an explanation of a physical phenomenon is not sufficiently clear unless it can be expressed in the terms of Newtonian mechanics.

It is here that the notion of *mechanism* emerges as the undisputed scientific paradigm. The *reductionist* trends which appear in all scientific domains (especially in chemistry and biology, but also in the social sciences) result from the imposition of this paradigm. They have determined the characteristics of the kind of scientific thinking accepted as valid until the twentieth century.

The type of paradigm that we are referring to is not imposed by socially accepted norms (as is the case for the choice of research topics), but constitutes the natural way to think about science at a particular period for any individual who takes an interest in science. There is no explicit external imposition. It is a conception that has become part of accepted knowledge and is transmitted along with it, as naturally as oral or written language is transmitted from one generation to the next. Hence we designate this kind of paradigm as an "epistemic paradigm" to distinguish it from the "social paradigm" described above.

This historical reminder is commonplace among those who have tried to analyze the influence of conceptions or "beliefs" on researchers at different times in history, and it is not likely to stir any controversy. However, the mechanisms by which the conceptions or beliefs held by a particular group (in the present case, the scientific community) act on the cognitive development of an individual has not been elucidated either by Kuhn or by any of the authors who have written about the subject of ideology and science. This epistemic paradigm is, however, the central topic of this chapter, since it is precisely the point of transition between the domain of the sociology to that of the sociogenesis of science.

What then is the mode of action of this epistemic paradigm? To answer this question, it will be necessary to refine the description of the paradigm given in the beginning of this chapter. As we noted above, any adult subject has already an elaborate arsenal of cognitive instruments enabling her to assimilate—and hence to interpret—the data she receives from the surrounding objects, as well as to assimilate the information transmitted to her by her society. This latter information refers to the objects and the situations already interpreted by the society in question. Following adolescence, when the fundamental logical structures that will constitute the basic instruments for her future cognitive development have been fully developed, the subject has at her disposal, in addition to these instruments, a conception of the world *(Weltanschauung)*, which determines her future assimilation of any experience. This conception of the world operates at several levels, and in a different way at each level, as we shall try to show in the following. But in order to make our thesis clear, we shall first present a historical example, which seems to us highly significant.

4. Ideology, Epistemic Framework and Paradigm

It is well known that when Greek civilization reached its apogee, science in China was developed to an extraordinary degree. The fundamental work of J. Needham, *Science and Civilisation in China,* is in this regard an inestimable source of documentation.[2]

A comparison between the characteristics of Greek and Chinese science is very enlightening in regard to the topic of the present chapter. In chapter 1, we saw that Aristotle (and all of mechanics up to Galileo) not only failed to succeed in formulating the principle of inertia, but in addition rejected as absurd any idea of a permanent movement not occasioned by the perpetual action of some force. Nevertheless, five centuries before Christ, we find the following statement made by a Chinese thinker: "The arrest of a movement is due to an opposing force. If there is no opposing force, the movement would never stop."

More than two millennia had to pass before Western science was to come to a similar conception. It is even more surprising that the statement just cited was not considered as an extraordinary discovery, but as a natural and obvious fact. The sentence that follows the citation in the Chinese text reads: "And this is as obvious as the fact that a cow is not a horse."

How is it to be explained that a statement that was absurd for the Greeks seemed obvious to the Chinese? We view this as one of the roots of the relationship between science and ideology. In addition, we believe that the answer to this question can throw light on one of the epistemological mechanisms by which the ideology of a society conditions the type of science that is going to develop in that society.

The Aristotelians conceived of the world as entirely static. The "natural state" of the objects of the physical world was, according to them, to be in a state of rest. All motion (except for the eternal motion of the stars upon which a divine force was thought to act) was considered as "violence" imparted on the object. Consequently, motion required the intervention of a force. When the force had stopped acting on the object, the object would return to its natural state of rest. It is obvious that in this kind of conception, the principle of inertia became *inconceivable.*

In contrast, for the Chinese the world was in a constant state of change. Movement, the continual flux, was the natural state of all things in the universe and therefore did not have to be explained. Force could intervene to modify or restrain movement. If no force acted on an object, the latter continued its motion without any change. The following may be the philosophical-religious foundation of this conception (Yand Hsing, 20 B.C.): "All things are generated by intrinsic impulses; only their weakening and their decline proceed in part from an external source."

It would be difficult to find a clearer example of the way two conceptions of the world (Weltanschauungen) lead to different physical explanations. The difference between one explanatory system and the other does not reside in any methodological difference nor in a different conception of science. It is an ideological difference which is reflected in a different epistemic framework, As a result, what is considered "absurd" or "obvious," relative to an epistemic framework is also partly determined by the dominant ideology. There does not seem to be any other way to explain the destiny of the principle of inertia in the Western world: absurd for the Greeks; a discovery of a truth inherent in the physical world for the seventeenth century; obvious and almost trivial for the nineteenth century (to the point where a student who would not "see" this principle as obvious would have been considered as seriously deficient); neither absurd nor obvious, neither true nor false for the twentieth century, since it is accepted only for the function it has in physical theory.

This static conception of the Greeks was one of the major obstacles (although it was not the only one they introduced) in the development of Western science. It was an ideological, not a scientific obstacle. The definite break with Aristotelian thinking, in the sixteenth and seventeenth centuries, was thus an ideological one, which lead to the introduction of a different epistemic framework and, in the end, to the imposition of a new epistemic paradigm.

Our interpretation is undoubtedly directly related to the position held by G. Bachelard, who was the first to have pointed to the importance of what he called "epistemological obstacle" and "epistemological rupture" in the development of science. We have already referred to this topic and indicated the points of similarity while also pointing out several differences, which we now have to discuss in more detail. In fact, Bachelard believes that there is a complete break between prescientific and scientific conceptions. At the same time, he identifies prescientific irrationalism as a major "epistemological obstacle." In our opinion, there is a greater continuity between prescientific and scientific thinking, to the extent that the mechanisms involved in the cognitive processes are the same. On the other hand, we also believe that there is a certain kind of "rupture" each time a transition takes place between one state of knowledge to another, in science as well as in psychogenesis. It seems reasonable to accept the idea of a rupture, if it is taken to mean a change in the epistemic framework.

In our view, at each moment in history, and in each society, there exists a dominant epistemic framework, a product of social paradigms, which in turn becomes the source of new epistemic paradigms. Once a given epistemic framework is constituted, it becomes impossible to dissociate the contribution of the social component from the one that is intrinsic to the cognitive system. That is, once it is constituted, the epistemic framework begins to act as an ideology which conditions the further development of science. This ideology functions as an epistemological obstacle preventing any development outside of the accepted conceptual framework. It is only at moments of crisis, of scientific revolutions, that there is a definite break with the dominant scientific ideology and that a different state results, characterized by a new epistemic framework, distinctly different from the preceding one.

Until now we have considered only one example showing the influence ideology can have on the conceptions of science. We shall consider other examples which show a different kind of relationship. Once again we shall use an example comparing Chinese and classical Greek thinking.

The Greek school of sophism is well known, even though it is often considered as some kind of historical curiosity, as one of the many surprising products of the Greek "miracle," without a parallel to other peoples of Antiquity. However, it is no simple historical accident that two other great civilizations of Antiquity, China and India, should have had their own schools of sophists at analogous periods of their history. In his thesis *Two Chinese sophists: Hounei Che Hů Shih and Kong-Sven Long,*[3] *Ignace Kuo Pao-Ku* points out that the political and social condition, at the time when the sophists made their appearance in China, were similar to those found in India and in Greece at the times when schools of sophists flourished there. In all three instances, these were periods when their civilizations had reached an apogee, and were followed by military and political disasters for the groups in power. This lead to a decay of the traditional institutions.

In China, this period corresponds to the disintegration of the feudal system and to the liquidation of its type of social organization. According to Fong-Yu-Lan, this was the origin of Chinese rationalism with its interest in logical argumentation and the appearance of the school of dialectics, out of which was born the Chinese school of sophists. In his *History of Chinese Philosophy,*[4] Fong-Yu-Lan characterizes this period as "a transitional period during which the institutions of the past had lost their hold, while those of the new era had not yet received their definitive formulation. Hence, this was inevitably a period of uncertainty and controversies." The period following the wars against the Medes in Greece presents the same characteristics. The same is true for the period at which flourished the negativist school of Buddhism (Ngarjuna, Aryaveda) in India.

In the three preceding examples, the demise of one historical order and its replacement by another, the dissolution of the social institutions and the break with "tradition," led to doubt concerning the foundations of the respective societies, including their beliefs and their accepted knowledge. In each case, one finds repeated assertions of the following kind: "Nothing can be affirmed"; "All truth is contradictory." For Nagarjuna, reality is *çunya,* that is, empty, but even to assert *çunya* (the void, nonexistence) is contradictory, and it too has to be denied. Reality thus becomes *çunya çunyata* (the void of the void), and so forth.

The Greeks showed that movement is impossible—that the arrow launched into the air is immobile, that Achilles could never catch up with the turtle. The Chinese showed that "a white horse is not a horse." The Indians demonstrated that nothing could "exist" or "not exist," or "exist and not exist," or "neither exist nor not exist."

In each case, the "established knowledge" is denied. In each case, this negation was erected into an ideology and thus stretched the logical analysis of knowledge to its ultimate limits.

We have here a different form of ideological influence, more directly inspired by the political and social facts than in the case cited previously, where the dominant element was a *Weltanschauung* of a philosophico-religious nature. Let us leave open the question concerning the origin of this *Weltanschauung,* which goes beyond the limits of the present book. It might be argued that it is an offspring of ideology, which, in the final analysis, is sociopolitical in nature.

The examples from Antiquity which we have cited (the principle of inertia and the sophist schools) were chosen because they all represent more or less "pure" cases—that is, situations in which the historical context is sufficiently clear so that it is possible to isolate to some extent the effect we intend to investigate. Generally, in historical development, the facts are not as clear nor is it generally possible to isolate such effects. Scientific progress, the search for new forms of explanation, the acceptance or the rejection of certain concepts or of a certain type of theory, most often are the result of a complex interplay of variables, in which social factors and the internal requirements of individual cognitive systems play complementary roles, reinforcing or opposing and attenuating one another.

II. THE CONTEMPORARY CONTROVERSY ABOUT THE DEVELOPMENT OF SCIENCE

How are the preceding considerations to be situated relative to the great controversy that was stirred up around the middle of the present century concerning the significance of scientific theories, specifically around Kuhn's position? Let us consider the latter in greater detail.

The notion of "paradigm" cannot be precisely defined within

Kuhn's theory. (One commentator found twenty-two different senses of the term in Kuhn's principal texts.) More recently, Kuhn himself acknowledged that he had used the term in more than one way, but he declared that all these different senses can be reduced to two basic ones. During a Symposium on the Structure of Scientific Theories Kuhn specified the two main uses of the term and assigned a specific name to each):

a. The "*disciplinary matrix*" is the paradigm in the strict sense. This type of paradigm could be defined as the "characteristic corpus of beliefs and of conceptions comprising all the *shared commitments* of a group of scientists."

b. "*Exemplars*": These are typical solutions to concrete problems that a group of scientists accepts as characteristic of the theory.[5]

This second sense of the word paradigm is important since, according to Kuhn, when a paradigm dominates the activity of a scientific community, it determines not only the theories and laws considered as valid, but also the type of problem and method of solution which are recognized as being scientific.

It is well known that Kuhn considered scientific evolution as being discontinuous. According to his conception, there are periods during which scientists work within a particular paradigm accepted by the community to which they belong (the disciplinary matrix); the basic principles of the paradigm are then not questioned: Students receive their training within the paradigm and learn how to solve the "paradigmatic" exemplars. The science they practice—which Kuhn calls "normal science"—consists in solving other problems that lie within the field of application of the theories constituting the paradigm. Their goal is what Kuhn called "*puzzle-solving*": there are problems which await solution and of which it is assumed that they *should* be solved with the theoretical instruments furnished by the accepted theories, even though they have not been solved until then. According to Kuhn, this activity is completely oriented by the conceptual limits of the paradigm in force so that a scientists's failure to solve one of these problems is considered as a personal failure, not as a failure of the theories used. When such failures occur repeatedly and a large domain of problems remains unsolved, in spite of repeated efforts of scientists whose competence is unquestionable, the theoretical framework itself begins to become the object of discussion. This leads to a crisis within the scientific discipline in question. Now, this does not mean that the theories are given up; new theories have to appear. A new paradigm capable of replacing the previous one has to be constituted (in the first sense of the term, the disciplinary matrix). This is the moment of a scientific revolution. The type of scientific research which leads to such a change in paradigm Kuhn called "revolutionary science" (in contradiction to what he called "normal science").

Thus, Kuhn characterized the development of science as a series of periods of no set duration during which "normal science" is practiced. Interrupting these periods are intervals of revolutionary science." Up to this point, we essentially agree with Kuhn. However, the epistemological implications he extrapolates from the history of science are completely different from our interpretations. The reason is probably that Kuhn takes history as the "memory of science" rather than as an "epistemological laboratory" (following the distinction made by Dijksterhuis, which we have used many times). Before we explain why we hold these divergent views, it is necessary to consider Kuhn's position in relation to other philosophers of science—regardless of their agreement with Kuhn's ideas.

Kuhn, who rigorously adhered to a line of reasoning advocated by Russell Hanson,[6] and also to some extent by Karl Popper,[7] affirmed, in opposition to logical empiricism, the impossibility of a neutral observational language—one independent of all theory. Hanson refers to this position as "the theory ladenness of all observational statements." According to Kuhn, scientists *observe the world* in a way that depends on the theories (paradigms) they accept as valid. True, the theories "correspond to the facts," but only because there is *information that has been transformed into facts*—information which did not exist for the preceding paradigms.

Kuhn evidently moves away from Popper's conception concerning the "plausibility" and the "refutation" of theories. He rejects Popper's stance on "refutation" by pointing out that several theories have been discarded before having been subjected to "test"; in addition, many theories have survived a long time after some of their assertions had been refuted by experiments. This leads to a second and more profound cleft between the views of Kuhn and Popper.

According to Popper, there exist certain well-defined criteria permitting scientists to make decisions between rival theories (plausibility and refutability). There exists, thus, a precise method that explains *progress* in science. "In science (and only in science), it is possible to say that there is real progress: that today we know more than was known previously." In contrast, Kuhn accepts neither the idea of continuity in the evolution of science, nor the idea that there exist well defined mechanisms for substituting one paradigm for another. Kuhn limits himself merely to noting a state of affairs when one paradigm gets historically replaced by another one. According to him, there are no applicable rules to explain how such a state of affairs comes about. Further, his conception of a paradigm makes it impossible to establish criteria for comparing and evaluating the respective validity of two different paradigms. At this point we encounter the "incommensurability" characteristic of paradigms, as Kuhn defined them. Many of Kuhn's critics see this as the source of Kuhnian irrationality, since in this conception there is no place for scientific progress.

Taking a similar stance (at least on a certain number of points), Feyerabend is ready to take this position to its final consequences.[8] Both Kuhn and Feyerabend agree that theories cannot be compared to experience (nor can they be refuted on the basis of experience), or to each other. Feyerabend, however, rejects the concept of "normal science," which he attributes to a monistic position. According to him, there is no absolute dominance of one paradigm (or theory), which is then replaced by another (the monistic position). Feyerabend defends a position which he calls *pluralistic,* and which he supports with examples from history: According to this position, at any given moment in history there coexist several theories. Sometimes they contradict each other, or even themselves. Individual scientists use one or the other according to their needs.

More recent among this group of philosophers of science is Imre Lakatos, who has continued Popper's line of reasoning. His goal is to avoid both the irrational consequences of Kuhn's position and the "epistemological anarchism" of the one put forth by Feyerabend.[9] In essence, Lakatos' thesis envisions the analysis of the dynamics of a science in terms of *sequences of interrelated theories,* rather than isolated theories. These "sequences" he calls "research programs." In a certain sense, the program takes the place of Kuhn's paradigm, adding to it Feyerabend's "pluralism." But, like Feyerabend and Popper, Lakatos rejects the concept of "normal science," which he considers a dogmatic and uncritical position, foreign to the attitude of a scientist.

Lakatos' main goal is to introduce into science Popper's type of rationality, which had been rejected by Kuhn and by Feyerabend.

In justifying his own epistemological position, Lakatos accepts Kuhn's criticism of Popper's theory of refutation ("falsificationism") as being partially correct: in fact, there is no such thing as "the crucial experiment" capable of immediately overthrowing a theory. Such an experiment is only one form of refutability (the dogmatic or "naïve" form). There exists a second form of refutability, which Lakatos attributes to Popper: "sophisticated falsificationism." Lakatos envisions his own contribution as being an elaborated version of this. He believes that this version avoids the pitfalls criticized by Kuhn.

Sophisticated falsificationism accomplishes a shift in the way the problems are formulated: instead of subjecting an isolated theory to empirical test it proposes to evaluate a *series of theories* (a research program). According to naïve falsificationism, a theory is refuted by an experiment (that is, a theory is falsified when a statement that expresses the result of an observation is in contradiction with a statement of the theory). According to sophisticated falsificationism, a theory is *refuted by another theory,* and not by an experiment.

In trying to compare the ideas of Popper, Kuhn, Feyerabend, and Lakatos, we can use either of two axes (dimensions) of reference, each represented by two alternatives:

a. *Rationality/irrationality.* Popper and Lakatos belong with the first of the alternatives; Feyerabend frankly belongs with the second and Kuhn similarly, but with certain caveats. Feyerabend's epistemological anarchism obliterates the distinction between "rational" and "irrational" poles. Kuhn claims to be able to guarantee a certain rationality on the basis of the rules of the game observed by the scientific community in the exercise of science. However, he presents neither the "rational mechanisms" of change in science nor the criteria to gauge progress. Popper and Lakatos seek rationality—and consequently they defend the notion of scientific progress—specifying criteria which make it possible to determine when and how a theory gets replaced by another one.

b. *Descriptive versus normative methodology.* Kuhn and Feyerabend attempt a description *ex post facto* of the way science proceeds and of what has made its historical development possible (even though their conclusion is that there is no "development" in the sense of a cumulative process resulting in "scientific progress"). Popper and Lakatos establish a heuristic—a methodology that establishes *norms.* Not only do they distinguish between science and pseudo-science, but they also formulate rules that scientists are obliged to respect.

It should be noted that all four authors make reference to the history of science to support their assertions. They draw different conclusions from the same history. Feyerabend's position is on the same side as Kuhn's with respect to the two dimensions just discussed. Nevertheless, Kuhn sees history as a sequence of paradigms, in which one paradigm remains dominant until it gets replaced by another, while Feyerabend has discovered the coexistence of conflicting paradigms, a proliferation of theories and contradictory hypotheses, of which only a few succeed in becoming preponderant. Popper and Lakatos also seem to be on the same side on the dichotomous dimensions we have distinguished. Yet, while Popper perceives in history a type of scientist who formulates adventurous conjectures and risky theories in order to subject these to the judgment of experience, at the risk of having to bury them if they are mortally wounded by observations contrary to their predictions, Lakatos perceives history as containing theories constituting "research programs"; in his view, theories are in constant confrontation with one another in order to establish which has sufficient merit to be preserved.

We should have added to our list of the four philosophers the names of N. Russell Hanson and of Stephen Toulmin. All six of them have the merit of having demonstrated the weaknesses of the neopositivist analysis of scientific knowledge. Since the 1950s (even though Popper had begun long before then), they have succeeded in demolishing the conception that had seen in scientific analysis nothing more than a process of rational reconstruction, entirely independent of the process of discovery. Aside from this commonality their positions diverge. They all agree that scientific analysis cannot be reduced to the *justification* of theories, but they differ with respect to the way they introduce the process of discovery. These differences then become very profound: according to the logical empiricists, there exists a "logic of justification" but not a "logic of discovery"; according to Popper, there is no logic of justification, but only a logic of discovery. According to Hanson, there is a logic of discovery and a logic of justification, but they are not the same.

III. THE NEED FOR AN EPISTEMOLOGICAL REFORMULATION

The situation just described is truly chaotic. Nevertheless, we believe that it is possible to introduce some order into this chaos, by using the elements of analysis we have proposed throughout. To do this, it will be necessary, first of all, to identify the problems, which in our view have been badly formulated. History shows that very often when scientists whose competence and degree of erudition are unquestionable arrive at very divergent points of view, something is amiss in the way the problem is formulated.

In the Introduction we mentioned some of the central points of support for our analysis. Our point of departure is that there is *continuity* in the development of the cognitive system, from the child to the average adult (one not educated in science) to the scientist. This continuity cannot be considered a postulate; it has to follow from ex post facto research. We believe that the comparative analysis of psychogenesis and the history of science presented in this book permits us to see the sense in which there is continuity as well as its limits. From Aristotelian and medieval physics (with their surprising similarities, both in method and in content, to the thought of children and adolescents) to the most highly developed branches of contemporary science (whose levels of abstraction are beyond the capacities of children and average adults) there exist mechanisms of action with strikingly common properties.

Now, this continuity in the regulatory mechanisms of cognitive development in no way excludes discontinuities in the process; on the contrary, it contributes to their determination. We have pointed out repeatedly that functional, not structural factors are really universal in cognitive development—both in the history of science and in psychogenesis. These have to do with the assimilation of novelties to existing structures and with their accommodation to new discoveries. The functional aspect of cognitive development explains the relative stability of the structures established, the process of disequilibration of a structure, and the process of re-equilibration of the system within a higher order structure. It is evident that the change from one structure to the next constitutes a *discontinuity,* a leap. It is also evident that such a change is neither predictable nor subject to norms. Given this kind of position, it is also possible to show that the structures thus elaborated have an internal stability which enables them to resist "perturbations" (for example, attempts at direct refutation). Up to this point, we agree with Kuhn (and also, partly, with Feyerabend) and disagree with Popper.

Still, we have to account for the fact that the changes in structure are not simply leaps in a vacuum; they follow an internal logic which has been documented in psychogenetic research for more than fifty years and sketched in the preceding chapters in our analyses of the way scientific theories become constituted. On this point we take a position which is opposed not only to that of Kuhn and Feyerabend, but also to the entire group we have cited. Even Popper and Lakatos, in trying to attribute rationality to the development of science and to defend the idea of scientific progress, limit themselves to the formulation of methodological norms designed to establish the acceptability of a theory or its rejection (Popper) or the criteria—also methodological—for evaluating the relative rank order of two or more theories (Lakatos). Both authors completely neglect the epistemological aspect of the problem. That problem consists, in our view, in establishing what precisely the change from a lower level theory T to the higher level theory T' consists in. This is a very different problem from the one formulated by Lakatos (given that for the latter, the problem consists in knowing how to establish that T' is superior to T).

We believe we have demonstrated some of these basic mechanisms in our study of the transitions from one of the great historical periods to the next, in the field of geometry, algebra and mechanics.

The conclusion we can draw from our analyses are somewhat odd. The neopositivists adopted an a priori position to the effect that the process of discovery is not relevant to what constitutes for them the basic goal of the philosophy of science: the justification of the validity of scientific knowledge. They did not seek any empirical basis for their theses, which they affirmed in the manner of dogmas, which certainly is in flagrant contradiction to their own principles. On the other hand, adherents of the school of Popper or Lakatos reject empiricism as a philosophical position with regard to the foundations of knowledge, but do not distinguish between the acceptance or rejection of empiricism as an "explanation" of the origin of knowledge, and the indisputable necessity to provide an empirical substratum to epistemological theses (in the absence of which one remains on the level of speculative philosophy). In Kuhn's case, the situation is even more troublesome, since he has tried to show how a student would learn his "exemplars."[10] He attempts to reconstruct the way a child learns what a duck is without taking the trouble to find out empirically (that is, by observing real children) whether this is really the way children learn. Several years of research with children have shown that children do not, in fact, learn the way Kuhn imagined. It is surprising to find Kuhn fall back on a neopositivist position—the very position he meant to demolish.

The refusal to use empirical research in order to determine how cognitive development really proceeds can be only explained by the presence of a preconceived idea, an a priori position concerning science—more precisely, the sciences. Such preconceived ideas are part of an ideology which determines the direction and the entire set of results of analysis.

We have tried throughout to provide support for the hypothesis, formulated in the Introduction, that there exists a certain functional continuity between the "natural," prescientific and the scientific subject (where the latter remains a "natural subject" outside of her scientific activity for as long as she does not defend a particular philosophical epistemology). If such a functional continuity exists, we can conclude that the two characteristics we attribute to all knowledge in the field of the sciences themselves are even more general than expected: the relative absence of conscious knowledge of its own mechanism and the continuously changing nature of the construction of knowledge. In fact, as the epistemological analysis of scientific thinking finds itself obliged to go back to its prerequisites, which are constituted by the cognitive elaborations of prescientific levels, this recursive procedure confronts us with increasingly unconscious structurings which are increasingly dependent upon their prior history. These two observations finally lead to a conclusion that is fundamental for an epistemology that claims to be objective rather than speculative—that in order to identify the origin of knowledge one has to follow a stepwise procedure all the way back to the level of actions.

Some comments are needed to put this formulation in perspective and to evaluate its scope. We have spoken of "the continuously changing nature of the construction of knowledge." But this is not meant to imply continuity in the mathematical sense of continuous function or curve. Let us repeat once again that this change can include breaks, leaps, disequilibria, and reequilibrations.

In addition, we have to insist on another aspect that is of decisive importance. When going back through the prescientific level all the way to the level of actions, we do not wish to suggest that one should consider the development of the subject facing objects that are already "given," independent of any social context. Rather, in the dialectical interaction between subject and object, the latter appears immersed within a network of relations.

After all that has been said, it may appear surprising that one can find similar models in the acquisition of knowledge repeated throughout history as well as psychogenesis at all levels. If society is to have so much influence, how is it that we find the same cognitive processes operating in all the different periods of human history and in all children, irrespective of social group and ethnicity? The answer is easy to find once we have identified the mechanisms to acquire knowledge that subjects have at their disposal and, parallel to that, the way the objects to be assimilated present themselves to the subject. Society can modify the latter, but not the former. The real meaning attributed to the object within the context of its interrelations with other objects may depend to a large extent on the way society acts upon the relations between subject and object. But the way this interpretation is acquired depends on the subject's cognitive mechanisms rather than on the contributions of the social group.

Because the subject's attention is directed toward certain objects (or situations) more than others, and because the objects are placed within certain contexts rather then others; the social environment and cultural models are very influential. However, all these conditions do not modify the mechanisms necessary for a biological species such as the human to acquire knowledge of these objects in their respective contexts as well as the particular type of social significance assigned in advance to these objects.

X

General Conclusions

At the end of our attempts to compare the history of science to the psychogenesis of knowledge, it may be useful to reexamine them by extracting the most general aspects and grouping these into three categories, which we shall call *instruments* common to all acquisition of knowledge, *processes,* which are the result of applying these instruments, and finally, *general mechanisms,* which represent a synthesis of the mechanisms by specifying their general direction.

I. INSTRUMENTS

The general source of the instruments of acquisition, which we have stressed only occasionally, is the *assimilation* of objects or events to the subject's existing schemata. This assimilation is virtually universal. It is true in all cases, from the infant's reflexes to the most highly developed forms of scientific thought. From a psychological perspective, assimilation is opposed to association, which is taken as a simple relation of similarity or contiguity between objects that are or will become known. This conception, according to which the subject's activity has no place in what comes to be known, and knowledge consists in nothing more than a heap of well categorized observables, is comparable to the contents of a box or a big wardrobe. From the point of view of science, such associationist empiricism is characteristic of positivism, which wants to reduce science to a series of "facts," which are first merely recorded and then described by means of a pure language constituted by a syntax and a semantics such as that developed in logic and mathematics. In contrast, assimilation characterizes knowledge as an indissociable relation between subject and object, where the latter constitutes a content upon which the subject imposes a form derived from her existing structures, adjusted for each new content; in particular, if the content is new, the assimilatory schemata are somewhat modified by means of *accommodations*—that is, differentiations in accordance with the object to be assimilated.

Now, this generality of assimilation, present in material form at the biological level (assimilation of food, of chlorophyl, etc.) and extending to functional forms at the cognitive level (sensorimotor, conceptual, etc.) has certain obvious implications for epistemology: the assimilatory nature of knowledge evidently contradicts not only all forms of empiricism (since it replaces the concept of knowledge as copy by that of knowledge as a continuous generation of structures) but also all forms of apriorism: while it is true that most of the biological forms of assimilation are hereditary, the essential characteristic of cognitive assimilation is to constantly construct new schemata from existing ones or to accommodate the latter. The assimilatory nature of knowledge thus demands a constructivist epistemology in the sense of a developmental or constructive structuralism, since to assimilate is to build structures. We have shown examples of this from history and from cognitive development. The main reason for the observed convergence in the two domains is precisely the fact that the subject plays an active part in the formation of knowledge and that the most general of her activities is assimilation.

As for the instruments of knowledge generated by assimilation, these are of course the generalizations and abstractions which have always been invoked by all forms of epistemology. However, assimilation enriches their traditional meaning, since it puts emphasis on the forms or schemata created by the subject as much as the contents it serves to structure. One distinction we have constantly utilized in our analyses is that between two forms of abstraction. We saw their alternate use in the constitution of physical knowledge, while only one form is involved in the progression of algebraic knowledge. The first of these forms may be called "empirical abstraction," in the sense that it applies to objects external to the subject, who notes certain of their properties in order to extract and analyze these. But, in physics and *a fortiori* in mathematics, we find a second form of abstraction, called "reflective abstraction," which applies to the subject's actions and operations as well as to the schemata which it leads the subject to construct. Now, we have pointed out repeatedly that "reflective" is to be understood in two different, but related ways: on the one hand, a "reflection" from a lower to a higher level (for example from action to representation) *versus* "reflexion" in the mental sense, as a reorganization, on a new level, of what is derived from the preceding one. In physics, we saw that there is constant alternation between empirical abstraction, applied to contents, and reflective abstraction, which derives from earlier forms the elements for new constructions, adapted to the new contents. In mathematics, however, where the subject elaborates both form and content (or if one so wishes, where everything is already "form" before it becomes "content"), reflective abstraction is naturally the only form of abstraction operative, in particular, in cases where an operation is first used instrumentally and then comes to be thematized, which permits the construction of a new theory.

Corresponding to these various forms of abstraction we find distinct forms of generalization. At the level of simple observation of empirical content, we find the extensional generalizations—that is, generalizations from "some" to "all," or from particular to general laws, without any reorganization of the former. In contrast, reflective abstraction allows the formation of completive and even constructive generalizations, which constitute new syntheses within which particular laws acquire new interpretations.

Let us further note that these notions of abstract reflective abstraction and constructive generalization and completive generalization lead to a specific interpretation of mathematics. Instead of considering mathematics as a system of deduction applied to "beings" given at the outset, be they quasi-empirical (cf. Gonseth's notion of "arbitrary object"), linguistic, or ideal (as in Platonism), we say that mathematics "derive from the subject's actions or operations." This means that (1.) the subject's action is always coordinated with other actions, since no action exists in isolation, their interpretation always being related to that of other actions; (2.) these coordinations are thus derived from forms detached from their contents; (3.) these forms, in turn, become coordinated with each other and constitute the origin of the basic operations, which, reflected, become the point of departure for logico-mathematical structures. To say that reflective abstraction "derives" its content from the subject's action is thus in no way a metaphor, but the expression of constructive activities present from the beginning, not pre-formed genetically nor simply recorded as a psychological observable, but normative and formative at all levels, including the most elementary ones.

As for the completive generalizations, they consist in going back from the whole to the parts, which get enriched by the addition of new knowledge. Many examples can be found of this in biology and physics, even though they are more common in mathematics. Thus, due to the electron theory of valences, Mendeleev's periodic table of elements, which had first been conceived of as a simple compilation of multiple measures, has become an instrument of new discoveries.

II. THE PROCESSES

The basic instruments of knowledge we have just enumerated give rise to various processes which it may be useful to briefly summarize. The most important of these is certainly the search for "reasons," which justify the abstractions and generalizations. Logical positivism has tried from its origin to get rid of this factor and to reduce science to a simple description of phenomena. This was A. Comte's idea. But in reality, every scientific mind, while not always admitting it, asks questions like that. It has often been noted that excellent physicists, while vigorously professing a positivist credo in the prefaces to their writings, contradict this faith in the body of their work by persuing a *bona fide* analysis of "causes"; as one example of this invincible tendency to search for "reasons," we might cite the evolution of contemporary mathematical logic. Limiting itself to a purely descriptive language, algebraic logic had long adhered to a purely extensional perspective, hence the "truth tables" which, in actuality, remain so far removed from any "truth" that they have led to the truly scandalous paradoxical situation that $p \supset q$ can be true when there is no actual relation of truth between p and q.

At present, we are witnessing the birth of a movement whose aims are to exclude all relations that are not logically necessary as well as significant so that each implication is based on a reason (cf. Anderson and Benlap's logic of *entailement*). Mathematicians, ever since Cournot, have distinguished between demonstrations which simply verify a theorem and those which, in addition, provide the reasons. Physicists today generally resort to explanatory "models," which clearly indicates that the generality of a law does not satisfy the mind and that an indomitable need exists for discovering reasons.

This general process which conforms to Leibniz's dictum *nil est sine ratione,* is accompanied by a correlative, showing (among other things) the role of the subject in knowledge: this is the procedure which consists in situating each real event between a class of possible events (e.g., the virtual effects) and a necessity conceived as being the only realizable possibility. In fact, neither the possible nor the necessary are observables, and both are the product of the subject's inferential activity. But since the subject is herself a part of reality by virtue of her status as a living organism, this enrichment resulting from her positioning herself between the possible and the necessary does not lead to epistemological idealism, but to a bipolar system, whose two poles only confirm the duality of form, due to the subject's assimilations, and of content, due to experience.

From this duality there results another, third fundamental process, which is the dual movement carrying assimilations and accommodations to a dynamic equilibrium between integrations and differentiations. In physics, it is evident that these two directions are the expression of the complex relation between a subject constantly approaching an object, which escapes comprehension as the discovery of new properties becoming available to knowledge raise ever new problems. But even in mathematics, it is no less obvious that the proactive invention of new structural systems is accompanied by retroactive differentiations introduced into subsystems whose lists had seemed to be exhausted.

This brings about another general process, which has become much more manifest during the more recent phases of mathematical constructivism than it had been in the platonism of preceding generations: this is the change from a phase in which certain operations play an instrumental role without becoming sufficiently conscious to a later phase, during which these same operations become thematized and thus lead to the development of new theories. (Cf. the "categories," functors, and morphisms thematized by McLane and Eilenberg, following their instrumental use in the elaboration of the Bourbakian "structures.")

Finally, the thematizations lead to yet another process that needs to be emphasized: the extension of the "reflective abstractions" (described above) to what we might call "reflexive abstractions," since they are the products of thematizations. A good example of this was discussed in our chapter on algebra with respect to the abstract "groups," which appeared after to the multiple particular groups known previously.

III. THE GENERAL MECHANISMS

It remains for us to summarize what we have said about the two main principal general mechanisms encountered again and again and which actually constitute a single mechanism in terms of their general significance: the transition from an intra to an inter and a trans phase on the one hand, and the general mechanism of equilibration on the other.

As for the former, which turned out to be the most general of all the commonalities between psychogenesis and the history of the sciences, it is easy to explain: The intra phase leads to the discovery of a set of properties in objects and events finding only local and particular explanations. The "reasons" to be established can thus be found only in the relations between objects, which means that they can be found only in "transformations." These, by their nature, are characteristic of the inter level. Once discovered, these transformations require the establishment of relations between each other, which leads to the construction of "structures," characteristic of the trans level.

Now, it is obvious that, as we have seen in particular in our chapter on algebra, the intra and inter levels, although achieving certain modes of equilibrium, generally give rise to many forms of disequilibrium and that the most finished forms of dynamic equilibrium are attained only by structures that have become stable as a result of connections established between transformations and as a result of interchange with the outside world of the kind characteristic of cognitive development (the integration of a limited structure within a larger (one) such that previous acquisitions are integrated within those taking their place.

To conclude, we wish to remind the reader that the constant aim of our genetic epistemology has been to show that the spontaneous development of knowledge has its source in biological organizations and tends toward the construction of logico-mathematical structures. We hope that the present volume, by demonstrating the role of psychogenesis and its remarkable convergence with the history of scientific thought, will help to lend support to this research program, even though we have not returned to the close relationship that one of us tried to establish on the basis of possible comparisons between biological and cognitive mechanisms.[1]

Now, Prigogine's recent work on "dissipative structures" seems to show that it is possible to go further and that the series "organism → behavior → sensorimotor → conceptual psychogenesis could be completed toward the lower end by relating the biological and hence cognitive structures to certain forms of dynamic equilibrium in physics (where the study of these structures was motivated precisely by the need to relate the two disciplines to each other).

In fact, there are at least five close analogies between these "dissipative structures" and what we consider equilibrations and cognitive equilibria. In the first place, these structures concern dynamic equilibria which include interchanges with the outside; these are quite different from equilibria without interchange. Secondly, it is these interchanges which stabilize the structures through regulations. Thirdly, equilibration as such is characterized, in both cases, by a form of "a self-organization." Fourthly, the states, at given points in time, having passed through a series of instable states, can be understood only on the basis of their past history.[2] Finally, and most importantly, the stability of a system is a function of its complexity. Thus, it is not surprising that Prigogine, in concluding the study cited here, can claim that his conception applies to a larger number of situations, "including the functioning of cognitive structures in J. Piaget's sense,"[3] and that, implicating "the observer, man in nature," it is in "complete agreement with the basic idea of genetic epistemology."

However, one difference remains, a difference which also sets the biological apart from the cognitive in general: in the case where one cognitive structure gets replaced by another, larger one, the old structure becomes integrated within the new one, which permits the continuity of knowledge, perfected in pure mathematics.

Notes

FOREWORD

1. Paris, PUF, 1950
2. Études d'épistémologie génétique vol. 14 (Paris, PUF, 1961).
3. Études d'épistémologie génétique, vol. 25 (Paris, PUF, 1971).
4. Études d'épistémologie génétique, vol. 26 (Paris, PUF, 1971).

INTRODUCTION

1. J. Piaget et al., *Recherches sur l'abstraction réfléchissante* (Paris; PUF, 1977), 2 vol. EEG 34 & 35.
2. We shall speak of "completive generalizations if a structure, while conserving its essential characteristics, becomes enriched with new subsystems, which are added without modifying the previous ones; for example, the noncommutative algebraic systems completing the commutative ones.
3. Beth, E. W. & Piaget, J. [1961] *Mathematical Epistemology and Psychology* (New York: Gordon & Breech, 1966).
4. Piaget, J. et al. *Recherches sur les correspondences* (Paris, PUF, 1980), EEG 37, chap. 4.
5. In his interesting "Introduction" to the book *Genèse de la pensée linguistique* (Paris, A. Colin, 1973), A. Jacob states that "an epistemological lesson of capital importance" has been for him to recognize that a reading of the development of linguistic theories favors neither continuity nor discontinuity" (p. 35).
6. J. Piaget & A. Szeminska (1941) *The Child's Conception of Number* (London: Routledge & Kegan Paul, 1952).

7. Piaget, J. & B. Inhelder [1962] *The Child's Construction of Quantities* (London: Routledge & Kegan Paul 1974).
8. B. Inhelder, A. Blanchet, A. Sinclair & J. Piaget, "Relations entre les conservations d'ensembles d'éléments discrets et celles de quantités continues" *Année psychologique* (1975) 75:23–60.
9. G. Juvet. *La structure des nouvelles theories physiques* (Paris, Alcan, 1933).
10. P. Greco, B. Inhelder, B. Matalon & J. Piaget. *La formation des raisonnements recurrentiels* (Paris, PUF, 1963), EEG 17.
11. P. Glansdorff & Prigogine, I. *Thermodynamic Theory of Structure, Stability and Fluctuations* (New York: Wiley, 1971), p. 62.
12. This dependence of the result on the path leading up to it is, of course, restricted to open systems. In contrast, a fundamental law of closed systems is that the same point can be reached by different paths. This is the case with groups in general and particularly the group of displacements. Another is that of categorical commutativity in the composition of functorial morphisms, etc.
13. J. Piaget et al. [1974b] *The Grasp of Consciousness: Action and Concept in the Young Child* (Cambridge: Harvard University Press, 1976), chapter 2.
14. H. Reichenbach, *Experience and Prediction* (Chicago, University of Chicago Press), 1938.
15. J. Piaget [1937] *The Construction of Reality in the Child* (New York: Basic Books, 1954).
16. J. Piaget. *L'Epistémologie génétique* (Paris, PUF, 1970), "Que sais-je?" no. 1399.

1. FROM ARISTOTLE TO THE MECHANICS OF THE IMPETUS

1. The citations from Aristotle's works are taken from the English translation by R. P. Hardie and R. K. Gaye (Physica) and J. L. Stooks (De Caelo) in the works of Aristotle, translated into English under the editorship of W. D. Ross, vol II. Oxford Clarendon Press, 1930. (trans.)
2. The citations from Philopon were taken from M. R. Cohen and I. E. Drabkin, *A Source Book of Greek Science* (New York, McGraw-Hill, 1948), pp. 221–223. All remaining citations were translated by the present translator directly from the French text. (trans.)
3. What is meant here is the earth as an element.
4. Cf. the extracts published by Cohen and Drabkin in *Source book of Greek Science.*
5. Erudissima commentaria in primos quatuor de naturali auscultatione libros, Venice, 1532. Extract translated by Duhem and reproduced by Dugas in his *Histoire de la mecanique,* (Neuchatel, Ed. du Griffon, 1950), p. 47.

6. A. C. Crombie, *Robert Grosseteste and the Origins of Experimental Science: 1100–1700* (Oxford, Clarendon Press, 1971) (reprint from 15th edition, 1953).
7. J. Buridan *Questions on the Eight Books of Aristotle's Physics.* The only text of Buridan's we have been able to consult concerning this point are the excerpts cited in de Clagett's book (Doc. 8,2).
8. N. Oresme, *Le livre du ciel et du monde* (Madison: The University of Wisconsin Press, 1968), p. 144.
9. In Marshall Clagett, ed. *Critical Problems in the History of Science* (Madison: The University of Wisconsin Press, 1969), pp. 103–104.
10. "Commentary on the Papers of A. C. Crombie and Joseph T. Clark" in Clagett, ed. *Critical Problems,* p. 141.
11. The Origin of Classical Mechanics from Aristotle to Newton" in Clagett, ed. *Critical Problems.*

2. PSYCHOGENESIS AND PRE-NEWTONIAN PHYSICS

1. J. Piaget. *The Child's Conception of Physical Causality* (London: Routledge & Kegan Paul, 1930). Note: (8;7) means 8 years and 7 months. Piaget's usual notation, which will be followed throughout this book.
2. J. Piaget et al. *La Transmission des mouvements* (Paris, 1972), EEG 27. J. Piaget and R. Garcia. *Understanding Causality.* New York Norton, 1974.
3. Piaget et al., *La Transmission des mouvements,* chap. 2.
4. J. Piaget et al., *La Formation de la notion de force* (Paris, PUF, 1973) EEG 29, chap. 2.
5. J. Piaget. "Le possible, l'impossible et le nécessaire." *Archives de psychologie* (1978) 44:281–299.
   J. Piaget. Essai sur la nécessité. *Archives de psychologie* (1977) 45(175):235–251.
   J. Piaget et al. *Possibility and Necessity* (Minneapolis: University of Minnesota Press PUF), vol. 1, *The Role of Possibility in Cognitive Development,* 1987; vol. 2, *The Role of Necessity in Cognitive Development,* 1987 (T). Translated from the French by H. Feider.
6. Some trivial examples: geometry, which was long considered as "necessarily Euclidean" or algebra, considered "necessarily commutative."
7. Piaget et al., *Possibility and Necessity,* vol. 1.
8. There have, in fact, been only a few applications, but very limited in number and only concerning secondary problems without a general epistemic framework.
9. B. Inhelder & J. Piaget. *The Growth of Logical Thinking from Childhood to Adolescence* (New York, Basic Books, 1958). H. Feider: (Minneapolis: University of Minnesota Press).

3. THE HISTORICAL DEVELOPMENT OF GEOMETRY

1. Poncelet, *Traité des Propriétiés Projectives des figures,* ouvrage utile a ceux qui s'occupent des applications de la Géometrie Descriptive et d'operations geometriques sur le terrain (Paris, Gauthier-Villars), 1865–1866.
2. Ibid., vol. 1, introduction, p. xi.
3. Ibid., vol. 1, pp. xii–xiii.
4. M. Chasles, *Aperçu historique, sur l'origine et le développement des Méthodes en Géometrie,* particulierement de celles qui se rapportent a la Géometrie Moderne, suivi d'une mémoire de Géometrie sur Deux Principes Généraux de la Science, la Dualité et l'Homographie (Paris: Gauthier-Villars, 1875).
5. Ibid., p. 95.
6. Poncelet, *Traité des Propriétes,* pp. xi–xii.
7. Italics as in the original.
8. Poncelet, *Traité des Propriétés,* pp. 207–208.
9. Chasles, *Aperçu historique,* p. 196.
10. Chasles, *Les Trois Livres des porismes d'Euclide* rétablis d'après la notice et les lemmes de Pappus (Paris, Mallet-Bacheleier, 1860), p. 14. Italics as in the original.
11. Chasles, *Les Trois Livres,* p. 18. Italics as in the original.
12. Pappus, *Mathematical Collections.* Translated from the Greek by Paul Ver Eecke (Paris et Bruges, Desclée de Brouwer), 1933.
13. Chasles, *Aperçu historique,* pp. 54–55. Italics as in the original.
14. J. Dieudonne, Preface to Felix Klein's *Programme d'Erlangen* (Paris, Gauthier-Villars, 1974), p. xi.
15. F. Klein, *Le Programme d'Erlangen* (Paris, Gauthier-Villars, 1974), p. 7.
16. Ibid.
17. On this position of Carnot's which was the source of his difficulties, and on its relation to Poncelet's position, see E. Bonipiani, "Il Principio di continuitá e l'imaginaraio in Géometria," *Questioni: riguardanti le matimatiche Elementari: Works in Honor of Federigo Enriques.*

4. THE PSYCHOGENESIS OF GEOMETRICAL STRUCTURES

1. For the experimental research, see J. Piaget, B. Inhelder, and A. Szeminska, *The Child's Conception of Geometry* (London: Routledge and Kegan Paul, 1960).
   J. Piaget & B. Inhelder, *The Child's Conception of Space* (London: Routledge & Kegan Paul, 1956).

2. J. Piaget, et al. *La Direction des mobiles lors de chocs et de poussées,* PUF, 1972 EEG 28, ch. 5.
3. J. Piaget et al. *Récherches sur la contradiction* (Paris, PUF, 1978 2 vols. EEG 31 & 32, vol. 2 chap. 11 Transl. *Experiments in contradiction* (Chicago: University of Chicago Press, 1980).
4. B. Inhelder, A. Blanchet, A. Sinclair, and J. Piaget "Relations entre les conservations d'ensembles d'éléments discrets et celles de quantités continues" *Année psychologique,* 75 (1975), pp. 23–60.
5. J. Piaget *The Child's Conception of Movement and Speed.* New York, Basic Books, 1970.
6. Piaget et al., *Recherches sur la généralisation* (Paris. PUF, 1978, chap. 9.
7. J. Piaget and C. Fot, Explication de la montée de l'eau dans un tube helicoidal, *Archives de psychologie,* 40 (1965), no 157 pp. 40–56.
8. J. Piaget et al., *Success and Understanding* (Cambridge: Harvard University Press, 1978).
9. V. Bank, et al., *L'Epistémologie de l'espace* (Paris: PUF, 1964), EEG 18.
10. J. Piaget et al., *La Composition des forces et le problème des vecteurs* (Paris: PUF, 1973), EEG 30.
11. We shall return to this point in the chapters on the psychogenesis of algebra.
12. J. Piaget et al., *Epistemology and Psychology of Functions* (Boston: Reidel, 1977).
13. J. Piaget, *Adaptation and Intelligence: Organic Selection and Phenocopy* (Chicago: University of Chicago Press, 1980).
14. J. Piaget, *Behavior and Evolution.* (New York: Pantheon Press, 1978).

5. ALGEBRA

1. J. Klein, *Greek Mathematical Thought and the Origin of Algebra* (Cambridge: MIT Press, 1968).
2. F. Vìete, *In Artem Analyticem Isagoge, Seorsim excusa ab opere restitutae Mathematicae Analyseos, seu, Algebra Nova,* Tours, 1591. Published in an English translation in J. Klein, *Greek Mathematical Thought.* The references to the passages cited in the following are taken from the following edition: F. Viete, *Opera Mathematica,* ed. by F. Shooten, Leyden, Elzevir, 1646. Translated into French by the author.
3. Vìete, *Opera Mathematica,* p. 1.
4. Ibid.
5. Ibid.
6. Ibid., p. 4.
7. Klein, *Greek Mathematical Thought,* p. 175.
8. Vìete, *Opera Mathematica,* p. 10.

9. Gauss, C. F. *Disquisitiones arithmeticae.* English translation by A. A. Clarke (New Haven: Yale University Press, 1966).
10. *Ecrits et mémoires mathématiques d'Evariste Galois* (edites par R. Bourgne and J. P. Azra) (Paris, Gauthier-Villars, 1962), p. 47; see also B. M. Kiernan, *The Development of Galois' Theory from Lagrange to Artin. Archive for the History of the Exact Sciences* (1971) 8:40–154.
11. Bourbaki, N. *Eléments d'histoire des mathématiques,* (Paris: Hermann, 1974), p. 104.
12. Cf. *Oeuvres mathématiques de Riemann,* (Paris: Albert Blanchard 1968). Klein's address was reproduced as a preface under the title "Riemann and His Influence on Modern Mathematics."
13. Dieudonne, *Cours de géométrie algébrique* (Paris, PUF) 1:42.
14. J. Piaget et al., (1978) Recherches sur la généralisation Presses Universitaires de France (Paris EEG, 1978) vol. 36.
15. Ibid., chap. 14.

6. THE FORMATION OF PRE-ALGEBRAIC SYSTEMS

1. J. Piaget, et al., *Recherches sur la généralisation* (Paris, PUF, 1978), EEG 36.
2. In the analytic and not the dialectic sense.
3. B. Inhelder and J. Piaget, *The Early Growth of Logic in the Child: Classification and Seriation* (New York: Harper & Row, 1964).

7. THE DEVELOPMENT OF MECHANICS

1. S. F. Mason, *A History of the Sciences* (New York: McMillan, 1962), p. 163.
2. Ibid., p. 162.
3. Ibid, p. 169.
4. A. Koyre, *Etudes galiléennes* (Paris, Herman, 1966), p. 164.
5. E. J. Dijksterhuis, "The Origins of Classical Mechanics from Aristotle to Newton," in M. Clagget ed., *Critical Problems in the History of Science* (Madison, University of Wisconsin Press, 1969), p. 175.
6. Cf. B. Ellis, "The Origin and Nature of Newton's Laws of Motion," in R. G. Cologny ed., *Beyond the Edge of Certainty: Essays in Contemporary Science and Philosophy* (Englewood Cliffs, New Jersey, Prentice Hall), 1965.
7. Cf. C. G. Pendse, "A Note on the Definition and Determination of Mass in Newtonian mechanics" *Philosophical Magazine* (1940) vol. 29.
8. Ibid.
9. Ibid.
10. J. L. Lagrange, *Mécanique analytique,* vol. 1, "la statique," first section.
11. Ibid., vol. 2, "La dynamique," first section.

12. Ibid.
13. Ibid., vol. 2, seventh section.
14. H. Poincaré, "Des fondements de la géométrie," *Revue de métaphysique et de morale* 1899.
15. Norwood Russell Hanson, *Patterns of Discovery* (Cambridge: Cambridge University Press, 1958).
16. The distinction between these two forms of abstraction has been established by developmental psychology (cf. the following chapter).

8. THE PSYCHOGENESIS OF PHYSICAL KNOWLEDGE

1. We shall return to these three phases in Section III (B).
2. J. Piaget et al. *Recherches sur l'abstraction réfléchissante* (Paris PUF, 1977), 2 vols. EEG 34 and 35.
3. J. Piaget and R. Garcia (1971) *Understanding Causality* (New York: Norton, 1974). For a detailed description of the experiments, see J. Piaget et al. *Travaux sur la causalité,* 3 vols. to appear.
4. Cf. the false translations at the level of semi-internal transmissions of movements.
5. Certainly, weight is observable and even its conservation can be understood empirically, but the pressure it exerts is not noticeable unless it produces a depression. Similarly, the fact that pressure can be felt when exerted on the palm of the hand does not imply, for young children, that such pressure also exists on hard objects.
6. J. Piaget (1946) *The Child's Conception of Movement and Speed* (New York: Basic Books, 1970).
7. Piaget and Garcia, *Understanding Causality.* For a detailed description of the experiments see Piaget et al. *Travaux sur la causalité.*
8. J. Piaget, (1946) *The Child's Conception of Time* (New York: Basic Books, 1970).
9. J. Piaget, et al. *Recherches sur la généralisation* (Paris, PUF, 1978) EEG 36, chap. 8.
10. H. E. Gruber, *Darwin on Man,* (New York, Dutton 1974).
11. It is well known that Descartes did not have a vectorial concept of speed (nor did Leibniz); consequently, the formula was erroneous even for the conservation of the quantity of motion.
12. J. Piaget et al. *Recherches sur l'abstraction réfléchissante,* vol. 2, ch. 12.
13. J. Piaget et al. *La Formation de la notion de force* (Paris PUF, 1973) EEG 29, chap. 5.
14. J. Piaget and R. Garcia, *Understanding Causality* (Paris, PUF, 1971), EEG 26.
15. J. Piaget & B. Inhelder (1962) *The Child's Construction of Quantities* (London: Routledge and Kegan Paul, 1974).

16. B. Inhelder and J. Piaget (1955) *The Growth of Logical Thinking from Childhood to Adolescence: An Essay on the Construction of Formal Operational Structures* (New York: Basic Books, 1958), chap. 2.
17. Piaget and Garcia, *Understanding Causality.* For a detailed description of the experiments, see Piaget et al. *Travaux sur la causalité.*
18. J. Piaget, *L'Equilibration des structures cognitives* (Paris, PUF, 1975), EEG 33, chap. 2, pp. 58–64.
19. For example, the decomposition of oblique lines into systems of orthogonal lines is at work virtually as soon as the subject conceives of these oblique lines as depending on two orientations simultaneously and reflective abstraction comes to make this explicit by dissociating the two.

9. SCIENCE, PSYCHOGENESIS, AND IDEOLOGY

1. One of the books in which he developed this approach was translated into French: Thomas Kuhn, *The Structure of Scientific Revolutions* (Paris, Flammarion, rev. ed. 1972; rev. ed. Champs, 1983).
2. A brief account in French can be found in Joseph Needham, *La Science Chinoise et l'Occident* (Paris, Seuil, 1978).
3. (Paris, PUF, 1953).
4. Fong-Yu-Lan, *A History of Chinese Philosophy* (Princeton, N.J. Princeton University Press, 1953).
5. Kuhn, "Second Thoughts on Paradigms." This text has been reproduced in F. Suppes, Ed., *The Structure of Scientific Theories,* (University of Illinois Press, 1977).
6. Norwood Russell Hanson, *Patterns of Discovery,* (Cambridge: Cambridge University Press, 1958).
7. K. Popper, *The Logic of Scientific Discovery* (London, Hutchinson, 1962).
8. P. Feyerabend, *Contre la méthode. Esquisse d'une théorie anarchiste de la connaissance* (Paris: Seuil, 1979).
9. I. Lakatos. "Falsification and Methodology of Scientific Research Programs" in I. Lakatos and A. Musgrave, eds., *Criticism and the Growth of Knowledge* (Cambridge: Cambridge University Press, 1970), p. 93.
10. Kuhn, *Second Thoughts on Paradigms.*

10. GENERAL CONCLUSIONS

1. J. Piaget, *Le Comportement, moteur de l'évolution* (Paris, Gallimard, 1976), Collection Idees no. 354.
2. I. Prigogine, "Physique et métaphysique," in *Connaissance scientifique et philosophie* (Bruxelles, Académie royale des sciences, 1975) p. 312.
3. Ibid., p. 316.

Index